D0886546

The mystery of the quantum world

The mystery of the quantum world

Euan Squires

*Department of Mathematical Sciences,
University of Durham*

Real are the dreams of Gods

John Keats
Lamia, **I**, 127

Adam Hilger Ltd
Bristol and Boston

British Library Cataloguing in Publication Data

Squires, Euan
   The mystery of the quantum world.
   1. Quantum theory
   I. Title
   530.1′2   QC174.12

   ISBN 0-85274-565-6
   ISBN 0-85274-566-4 Pbk

Published by Adam Hilger Ltd
Techno House, Redcliffe Way, Bristol BS1 6NX, England
PO Box 230, Accord, MA 02018, USA

Typeset by Mathematical Composition Setters Ltd, Salisbury
Printed in Great Britain by J W Arrowsmith Ltd, Bristol

Contents

Preface

This book is about physics. Within the limitations of our present knowledge, I believe that what it says about physics is correct. More particularly, it is about quantum physics; about the mysterious behaviour of the micro-world, and the strange properties of the quantum theory which predicts this behaviour. In an endeavour to understand the quantum world, we are led beyond physics, certainly into philosophy and maybe even into cosmology, psychology and theology. Here I am not sure that there are clear criteria of what it means to be 'correct', and even if there are, I have less confidence that what is said will always satisfy these criteria. I have ventured into these areas because the issues raised by quantum physics are relevant, important and interesting.

Quantum phenomena challenge our primitive understanding of reality; they force us to re-examine what the concept of existence means. These things are important, because our belief about what *is* must affect how we see our place within it, and our belief about what we are. In turn, what we *believe* we are ultimately affects what we *actually* are and, therefore, how we behave. Nobody should ignore physics.

E J Squires

Acknowledgments

I have had useful discussions with many people during the writing of this book. In particular, I would like to thank Peter Collins, Peter Rowe, Barry Gower and Timothy Squires, all of whose comments on earlier attempts have contributed significantly to the final version.

Chapter One

Reality in the Quantum World

1.1 The quantum revolutions

Quantum mechanics, created early this century in response to certain experimental facts which were inexplicable according to previously held ideas (conveniently summarised by the title 'classical physics'), caused three great revolutions. In the first place it opened up a completely new range of phenomena to which the methods of physics could be applied: the properties of atoms and molecules, the complex world of chemical interactions, previously regarded as things given from outside science, became calculable in terms of a few fixed parameters. The effect of this revolution has continued successfully through the physics of atomic nuclei, of radioactivity and nuclear reactions, of solid-state properties, to recent spectacular progress in the study of elementary particles. In consequence all sciences, from cosmology to biology, are, at their most fundamental level, branches of physics. Through physics they can, at least in principle, be understood. Indeed, on contemplating the success of physics, it is easy to be seduced into the belief that 'everything' is physics—a belief that, if it is intended to imply that everything is understood, is certainly false, since, as we shall see, the very foundation of contemporary theoretical physics is mysterious and incomprehensible.

The second revolution was the apparent breakdown of *determinism*, which had always been an unquestioned ingredient and an inescapable prediction of classical physics. Note that we are using the word 'determinism' solely with regard to physical systems, without at this stage worrying about which systems can be so described; that is, we are not here concerned with such concepts as free will. In a deterministic theory the future behaviour of an isolated physical system is uniquely determined by its present state. If, however, the world is correctly described by quantum theory, then, even for simple systems, this deterministic property is not valid. The outcome of any particular experiment is not, even in principle, predictable, but is chosen at random from a set of possibilities; all that can be predicted is the probability of particular results when the experiment is repeated many times. It is important to realise that the probability aspects that enter here do so for a different reason than, for example, in the tossing of a coin, or throw of a dice, or a horse race; in these cases they enter because of our lack of precise knowledge of the orginal state of the system, whereas in quantum theory, even if we had complete knowledge of the initial state, the outcome would still only be given as a probability.

Naturally, physicists were reluctant to accept this breakdown of a cherished dogma—Einstein's objection to the idea of God playing dice with the universe is the most familiar expression of this reluctance—and it was suggested that the apparent failure of determinism in the theory was due to an incompleteness in the description of the system. Many attempts to remedy this incompleteness, by introducing what are referred to as 'hidden variables', have been made. These attempts will form an important part of our later discussion.

We are accustomed to regarding the behaviour, at least of simple mechanical systems, as being completely deterministic, so if the breakdown of determinism implied by quantum mechanics is genuine, it is an important discovery which must affect our view of the physical world. Nevertheless, our belief in determinism arises from experience rather than logic, and it is quite possible to conceive of a certain degree of randomness entering into mechanics; no obvious violation of 'common sense' is involved. Such is not the case with the third revolution brought about by quantum mechanics. This challenged the basic belief, implicit in all science and indeed in almost the whole of human thinking, that there exists an objective reality, a reality that does not depend for its existence on its being observed. It is because of this challenge that all who endeavour to study, or even take an interest in, reality, the nature of 'what is', be they philosophers or theologians or scientists, unless they are content to study a phantom world of their own creation, should know about this third revolution.

To provide such knowledge, in a form accessible to non-scientists, is the aim of this book. It is not intended for those who wish to learn the practical aspects of quantum mechanics. Many excellent books exist to cover such topics; they convincingly demonstrate the power and success of the theory to make correct predictions of a wide range of observed phenomena. Normally these books make little reference to this third revolution; they omit to mention that, at its very heart, quantum mechanics is totally inexplicable. For their purpose this omission is reasonable because such considerations are not relevant to the success of quantum mechanics and do not necessarily cast doubt on its validity. In 1912, Einstein wrote to a friend, 'The more success the quantum theory has, the sillier it looks.' [Letter to H Zangger, quoted on p 399 of the book *Subtle is the Lord* by A Pais (Oxford: Clarendon 1982).] If it is true that quantum mechanics is 'silly', then it is so because, in the terms with which we are capable of thinking, the world appears to be silly. Indeed the recent upsurge of interest in the topic of this book has arisen from the results of recent experiments; results which, though they beautifully confirm the predictions of quantum mechanics, are themselves, quite independent of any specific theory, at variance with what an apparently convincing, common-sense, argument would predict (see Chapter 5, especially §§5.4 and 5.5, for a complete discussion of these results).

We can emphasise the essentially observational nature of the problem we are discussing by returning to the experimental facts we mentioned at the start of this section, and which gave birth to quantum mechanics. Although, by abandoning some of the principles of classical physics, quantum theory *predicted* these facts, it did not *explain* them. The search for an explanation has continued and we shall endeavour in this book to outline the various possibilities. *All involve radical departures from our normal ways of thinking about reality.*

On almost all the topics which we shall discuss below there is a large literature. However, since this book is intended to be a popular introduction rather than a technical treatise, I have given very few references in the text but have, instead, added a detailed bibliography. For the same reason various *ifs* and *buts* and qualifying clauses, that experts might have wished to see inserted at various stages, have been omitted. I hope that these omissions do not significantly distort the argument.

I have tried to keep the discussion simple and non-technical, partly because only in this way can the ideas be communicated to non-experts, but also because of a belief that the basic issues are simple and that highly elaborate and symbolic treatments only serve to confuse them, or, even worse, give the impression that problems have been solved when, in fact, they have merely been hidden. The appendices, most of which require a little more knowledge of mathematics and physics than the main text, give further details of certain interesting topics.

Finally, I conclude this section with a confession. For over thirty years I have used quantum mechanics in the belief that the problems discussed in this book were of no great interest and could, in any case, be sorted out with a few hours careful thought. I think this attitude is shared by most who learned the subject when I did, or later. Maybe we were influenced by remarks like that with which Max Born concluded his marvellous book on modern physics [*Atomic Physics* (London: Blackie 1935)]: 'For what lies within the limits is knowable, and will become known; it is the world of experience, wide, rich enough in changing hues and patterns to allure us to explore it in all directions. What lies beyond, the dry tracts of metaphysics, we willingly leave to speculative philosophy.' It was only when, in the course of writing a book on elementary particles, I found it necessary to do this sorting out, that I discovered how far from the truth such an attitude really is. The present book has arisen from my attempts to understand things that I mistakenly thought I already understood, to venture, if you like, into 'speculative philosophy', and to discover what progress has been made in the task of incorporating the strange phenomena of the quantum world into a rational and convincing picture of reality.

1.2 External reality

As I look around the room where I am now sitting I *see* various objects. That is, through the lenses in my eyes, through the structure of the retina, through assorted electrical impulses received in my brain, etc, I experience sensations of colour and shape which I interpret as being caused by objects outside myself. These objects form part of what I call the 'real world' or the 'external reality'. That such a reality exists, independent from my observation of it, is an assumption. The only reality that I *know* is the sensations of which I am conscious, so I make an assumption when I introduce the concept that there are real external objects that cause these sensations. Logically there is no need for me to do this; my conscious mind could be all that there is. Many philosophers and schools of philosophy have, indeed, tried to take this point very seriously either by denying the existence of an external reality, or by claiming that, since the concept cannot be properly defined, proved to exist, or proved not to exist, then it is useless and should not be discussed. Such views, which as philosophic theories are referred to by words such as 'idealism' or 'positivism', are logically tenable, but are surely unacceptable on aesthetic grounds. It is much easier for me to understand my observations if they refer to a real world, which exist even when not observed, than if the observations are in fact everything. Thus, we all have an intuitive feeling that 'out there' a real world exists and that its existence does not depend upon us. We can observe it, interact with it, even change it, but we cannot make it go away by not looking at it. Although we can give no proof, we do not really doubt that 'full many a flower is born to blush unseen, and waste its sweetness on the desert air'.

It is important that we should try to understand why we have this confidence in the existence of an external reality. Presumably one reason lies in selective evolution which has built into our genetic make-up a predisposition towards this view. It is easy to see why a tendency to think in terms of an external reality is favourable to survival. The man who sees a tree, and goes on to the idea that there *is* a tree, is more likely to avoid running into it, and thereby killing himself, than the man who merely regards the sensation of seeing as something wholly contained within his mind. The fact of the built-in prejudice is evidence that the idea is at least 'useful'. However, since we are, to some extent, thinking beings, we should be able to find rational arguments which justify our belief, and indeed there are several. These depend on those aspects of our experience which are naturally understood by the existence of an external reality and which do not have any natural explanation without it. If, for example, I close my eyes and, for a time, cease to observe the objects in the room, then, on reopening them, I see, in general, the same objects. This is exactly what would be expected on the assumption that the objects exist and are present even when I do not actually look at them. Of course, some could have moved, or even been taken away, but in this case I would seek, and normally find, an explanation of the changes. Alternatively I could use different methods of 'observing', e.g. touch, smell, etc, and I would find that the same set of objects, existing in an external world, would explain the new observations. Thirdly, I am aware through my consciousness of other people. They appear to be similar to me, and to react in similar ways, so, from the existence of my conscious mind, I can reasonably infer the existence of real people, distinct from myself, also with conscious minds. Finally, these other people can communicate to me their observations, i.e. the experiences of their conscious minds, and these observations will in general be compatible with the same reality that explains my own observations.

In summary, it is the *consistency* of a vast range of different types of observation that provides the overwhelming amount of evidence on which we support our belief in the existence of an external reality behind those observations. We can contrast this with the situation that occurs in hallucinations, dreams, etc, where the lack of such a consistency makes us cautious about assuming that these refer to a real world.

We turn now to the scientific view of the world. At least prior to the onset of quantum phenomena this is not only consistent with, but also implicitly assumes, the existence of an external reality. Indeed, science can be regarded as the continuation of the process, discussed above, whereby we explain the experiences of our senses in terms of the behaviour of external objects. We have learned how to observe the world, in ever more precise detail, how to classify and correlate the various observations and then how to explain them as being caused by a real world behaving according to certain laws. These laws have been deduced from our experience, and their ability to predict new phenomena, as evidenced by the enormous success of science and technology, provides impressive support for their validity and for the picture of reality which they present.

This beautifully consistent picture is destroyed by quantum phenomena. Here, we are amazed to find that one item, crucial to the whole idea of an external reality, appears to fail. It is no longer true that different methods of observation give results that are consistent with such a reality, or at least not with a reality of the form that had previously been assumed. No reconciliation of the results with an acceptable reality has been found. This is the major revolution of quantum theory, and, although of no immediate practical importance, it is one of the most significant discoveries of science and nobody who studies the nature of reality should ignore it.

It will be asked at this stage why such an important fact is not immediately evident and well known. (Presumably if it had been then the idea of creating a picture of an external reality would not have arisen so readily.) The reason is that, on the scale of magnitudes to which we are accustomed, the new, quantum effects are too small to be noticed. We shall see examples of this later, but the essential point is that the basic parameter of quantum mechanics, normally denoted by $\hbar$ ('h bar') has the value $0.000\,000\,000\,000\,000\,000\,000\,000\,001$ (approximately) when measured in units such that masses are in grams, lengths in centimetres and times in seconds. (Within factors of a thousand or so, either way, these units represent the scale of normal experience.) There is no doubt that the smallness of this parameter is partially responsible for our difficulty in understanding quantum phenomena—our thought processes have been developed in situations where such phenomena produce effects that are too small to be noticed, too insignificant for us to have to take them into account when we describe our experiences.

1.3 The potential barrier and the breakdown of determinism

We now want to describe a set of simple experiments which demonstrate the crucial features of quantum phenomena. To begin we suppose that we have a flat table on which there is a smooth 'hill'. This is illustrated in figure 1. If we roll a small ball, from the right, towards the hill then, for low initial velocities, the ball will roll up the hill, slowing down as it does so, until it stops and then rolls back down again. In this case we say that the ball has been *reflected*. For larger velocities, however, the ball will go right over the hill and will roll down the other side; it will have been *transmitted*.

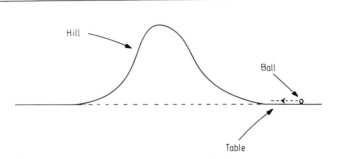

Figure 1 A simple example of a potential barrier experiment, in which a ball is rolled up a hill. The ball will be reflected or transmitted by the hill according to whether the initial velocity is less or greater than some critical value.

By repeating this experiment several times we readily find that there is a critical velocity, which we shall call V, such that, if the initial velocity is smaller than V then the ball will be reflected, whereas if it is greater than V then it will be transmitted. We can write this symbolically as

$$v < V: \text{reflection}$$
$$v > V: \text{transmission} \tag{1.1}$$

where v denotes the initial velocity, and the symbols $<$, $>$ mean 'is less than', 'is greater than', respectively.

 The force that causes the ball to slow down as it rises up the hill is the gravitational force, and it is possible to calculate V from the laws of classical physics (details are given in Appendix 1). Similar results would be obtained with any other type of force. What is actually happening is that the energy of motion of the ball (called kinetic energy) is being changed into energy due to the force (called potential energy). The ball will have slowed to zero velocity when all the kinetic energy has turned into potential energy. Transmission happens when the initial kinetic energy is greater than the maximum possible potential energy, which occurs at the top of the hill. In the general case we shall refer to this type of experiment as reflection or transmission by a *potential barrier*.

Now we introduce quantum physics. The simple result expressed by equation (1.1), which we obtained from experiment and which is in agreement with the laws of classical mechanics, is not in fact correct. For example, even when $v < V$ there is a possiblity that the particle will pass through the barrier. This phenomenon is some-times referred to as *quantum tunnelling*. The reason why we would not see it in our simple laboratory experiment is that with objects of normal sizes (which we shall refer to as 'macroscopic' objects), i.e. things we can hold and see, the effect is far too small to be noticed. Whenever v is measurably smaller than V the probability of transmission is so small that we can effectively say it will never happen. (Some appropriate numbers are given in Appendix 4.)

With 'microscopic' objects, i.e. those with atomic sizes and smaller, the situation is very different and equation (1.1) does not describe the results except for sufficiently small, or sufficiently large, velocities. For velocities close to V we find, to our surprise, that the value of v does not tell us whether or not the particle will be transmitted. If we repeat the experiment several times, always with a fixed initial velocity (v) we would find that in some cases the particle is reflected and in some it is transmitted. The value of v would no longer determine precisely the fate of the particle when it hits the barrier; rather it would tell us the *probability* of a particle of that velocity passing through. For low velocities the probability would be close to zero, and we would effectively be in the classical situation; as the velocity rose towards V the probability of transmission would rise steadily, eventually becoming very close to unity for v much larger than V, thus again giving the classical result.

Before we comment on the implications of these results, it is worth considering a more readily appreciated situation which is in some ways analogous. On one of the jetties in the lake of Geneva there is a large fountain, the 'Jet d'eau'. The water from this tends to fall onto the jetty, in amounts that vary with the direction of the wind. On any day in summer people walk along the jetty and eventually they reach the 'barrier' of the falling water. At this stage some are 'reflected', they look around for a while and then turn back; others however are 'transmitted' and, ignoring the possibility of getting wet, carry on to the end of the jetty. By observing for a time, on any particular afternoon, it would be possible to calculate the probability that any given person would pass the barrier. This probability would depend on the direction of the wind at the time of observation—the direction would therefore play an analoguous role to that of the initial velocity in our previous experiment. There would, however, be nothing in any way surprising about our observations at Geneva, no breakdown of determinism would be involved, people would behave differently because they *are* different. Indeed it might be possible to predict some of the effects: the better dressed, the elderly, the female (?)... would, perhaps, be more likely to be reflected. The more information we had, the better would we be able to predict what would happen and, indeed, leaving aside for the moment subtle questions about free will which inevitably arise because we are discussing the behaviour of people, we might expect that if we knew everything about the individuals we could say with certainty whether or not they would pass the barrier. In this sense the probability aspects would arise solely from our ignorance of all the facts—they would not be intrinsic to the system. In all cases where probability enters classical physics this is the situation.

We must contrast this perfectly natural happening with the potential barrier experiment. Here the particles are, apparently, identical. What then determines which are reflected and which transmitted? Attempts to answer this question fall into two classes:

Orthodox theories. In such theories it is accepted that the particles genuinely are identical, so there is nothing available with which to answer the question except the statement that it is a random choice, subject only to the requirement that when the same experiment is repeated many times the correct proportion have been reflected. Quantum theory, as normally understood, is a theory of this type. If such theories are correct then determinism, as defined in §1.1, is not a property of our world; probability enters physics in an intrinsic way and not just through our ignorance. The situation is thus different in nature from that of people passing the Jet d'eau in Geneva. Herein lies the second revolution of quantum physics to which we referred in the opening section. The physical world is not deterministic. It is worth noting here that, although quantum phenomena are readily seen only on the microscopic scale, this lack of determinism can easily manifest itself on any macroscopic scale one might choose. We give a simple example in Appendix 2.

Hidden variable theories. In such theories the particles reaching the barrier are not identical; they possess other variables in addition to their velocities and, in principle, the values of these variables determine the fate of each particle as it reaches the barrier; no breakdown of determinism is required and the probability aspect only enters through our ignorance of these values, exactly as in classical physics. At this stage of our discussion readers are probably thinking that hidden variable theories surely contain the truth, and that we have not yet given any good reasons for abandoning determinism. They are right, but this will soon change and we shall see that hidden variable theories, which are discussed more fully in Chapter 5, have many difficulties.

Before proceeding we shall look a little more carefully at our potential barrier experiment. Since we are interested in whether or not particles pass through the barrier we must have detectors which record the passage of a particle, e.g. by flashing so that we can see the flash. We shall assume that our detectors are 'perfect', i.e. they never miss a particle. Then if we have a detector on the left of the barrier it will flash when a particle is transmitted, whereas one on the right will flash for a reflected particle. Suppose N particles, all with the same velocity, are sent and suppose we see R flashes in the right-hand detector and T in the left-hand detector. Because every particle must go somewhere, we will find

$$R + T = N. \tag{1.2}$$

Provided N is large, the probability of transmission is defined to be T divided by N and the probability of reflection R divided by N, i.e.

$$P_T = T/N \tag{1.3}$$

and

$$P_R = R/N \tag{1.4}$$

where P_T and P_R denote the probabilities of transmission and relfection, respectively.

If we were to repeat the experiments, using N further particles, then we would not obtain exactly the same values for R and T. (Compare the fact that in 100 tosses of a coin we would not always obtain exactly 50 heads.) These differences are statistical fluctuations and their effect on the values of P_T and P_R can be made as small as we desire by making N large enough. In fact, the error is proportional to the inverse of the square root of N. In all the subsequent discussion we shall assume that N is sufficiently large for statistical fluctuations to be ignored.

At this stage everything in our experiment appears to be in accordance with the concept of external reality. Indeed we have a simple picture of what happens: each particle moves freely until it reaches the potential barrier, at which stage it makes a 'choice', either through a hidden variable procedure or with some degree of randomness, as to whether to pass through or not. Such a choice would be made regardless of whether the detectors were present. After a suitable lapse of time we would have either a particle travelling to the right or one travelling to the left. This would be the external reality. If the detectors were present one of them would flash, thereby telling us which of the two possibilities had occurred. The detectors however would only *observe* the reality, they would not *create* it.

This simple picture of reality is, as we shall now show, false. It is not compatible with another method of observing the same system and therefore fails one of the consistency tests for reality given in §1.2. In the next section we shall describe this other method of observation and see why it is so devastating to the idea of external reality.

1.4   The experimental challenge to reality

We continue with our experiment in which particles are directed at a potential barrier but now, instead of having detectors to tell us whether a particle has been reflected or transmitted, we have 'mirrors' which deflect both sets of particles towards a common detector. There are many ways of constructing such mirrors, particularly if our particles are charged, e.g. if they are electrons, when we could use suitable electric fields. For this experiment we must also allow the particles to follow slightly different paths, which can easily be arranged if there is some degree of variation in the initial direction. To be specific, we suppose that the source of particles gives a uniform distribution over some small angle. Then the final detector must cover a region of space sufficiently large to see particles following all possible paths. In fact, we split it into several detectors, denoted by A, B, C, etc, so that we will be able to observe how the particles are distributed among them. In figure 2 we give a plan of the experiment. This plan also shows two separate particle paths reaching the detector labelled C.

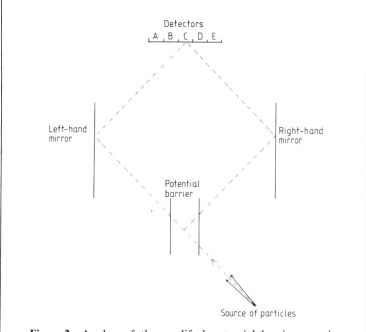

Figure 2 A plan of the modified potential barrier experiment. The mirrors can be put in place to deflect the reflected and transmitted particles to a common set of detectors. Two possible particle paths to detector C are shown.

We now do three separate sets of experiments. For the first set we only have the right-hand mirror. Thus only the particles that are reflected by the barrier will be able to reach the detectors. When we have sent N particles, where N is large, the detectors will have flashed R times. These R flashes will have some particular distribution among the various detectors. A possible example of such a distribution, for five detectors, is shown in figure 3 (*a*).

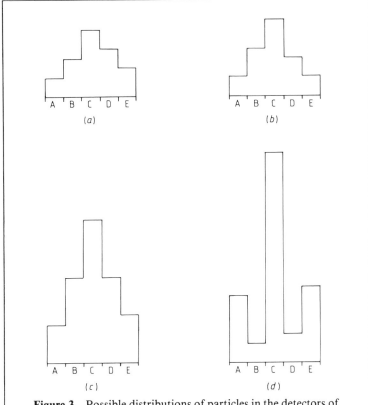

Figure 3 Possible distributions of particles in the detectors of figure 2. The height above each letter indicates the number of particles received by the corresponding detector. In (*a*) only the right-hand mirror is present; in (*b*) only the left-hand mirror. In (*c*) we show the sum of (*a*) and (*b*), while in (*d*) we show the actual distribution when both mirrors are in position.

Next, we repeat these experiments with the right-hand mirror removed and the left-hand mirror in place. This time only the transmitted particles will reach the detectors, so, when we have sent N particles, we will have T flashes. In figure 3 (b) we show a possible distribution of these among the same five detectors.

For our third set of experiments we have both mirrors in position. Thus all particles, whether reflected or transmitted by the barrier, will be detected. When N particles have been sent, there will have been N flashes. Can we predict the distribution of these among the various detectors? Surely, we can. We know what happens to the transmitted particles, e.g. figure 3 (b), and also to the reflected particles, e.g. figure 3 (a). We also know that the particles are sent separately so they cannot collide or otherwise get in each other's way. We therefore expect to obtain the sum of the two previous distributions. This is shown in figure 3 (c) for our example. The world, however, is not in accord with this expectation. The distribution seen when both mirrors are present is not the sum of the distributions seen with the two mirrors separately. Indeed, it is quite possible for some detectors to receive fewer particles when both mirrors are present than when either one is present. A typical possible form showing this effect is given in figure 3 (d).

Can we understand these results? Can we understand, for example, why there are paths for particles to reach detector B when either mirror is present but such paths are not available if both mirrors are present? The only possibility is that in the latter case each individual particle 'knows about', i.e. is influenced by, both mirrors. This is not compatible with the view of reality, discussed in the previous section, in which a particle either passes through or is reflected. On the contrary, the reality suggested by the experiments of this section is that each particle somehow splits into two parts, one of which is reflected by one mirror and one by the other. Such a picture is, however, not compatible with the results of the detector experiments in which each individual particle is seen to go one way or the other and never to split into two particles. Thus the simple pictures of reality suggested by these two sets of experiments are mutually contradictory.

Clearly we should not accept this perplexing situation without examing very carefully the steps that have led to it. The first thing we would want to check is that the experimental results are valid.

Here I have to make an apology. Contrary to what has been implied in the above discussion, the experiments that have been described have not actually been done. For a variety of technical reasons no real experiment can ever be made quite as simple as a 'thought' experiment. The apparent incompatibility we have met does occur in real experiments, but the discussion there would be much more complicated and the essential features would be harder to see. The 'results' of our simple experiments actually come from theory, in particular from quantum theory, but the success of that theory in more complicated, real, situations means that we need have no doubt about regarding them as valid experimental results.

As another possibility for rescuing the picture of reality given in the previous section, we might ask whether we abandoned it too readily in the face of the evidence from the mirror experiments. On examining the argument we see that a key step lay in the statement that a reflected particle, for example, could not know about the left-hand mirror. Behind this statement lay the assumption that objects sufficiently separated in space cannot influence each other. Is this assumption true and, if so, were our mirrors sufficiently well separated? With regard to the second question one answer is that, according to quantum mechanics, which provided our results, the distance is irrelevant. Perhaps more important, however, is the fact that the irrelevance of the distance scale seems to be experimentally supported in other situations. The only hope here, then, is to question the assumption; maybe the belief that objects can be spatially separated so that they no longer influence each other is false. If this is so, then it is already a serious criticism of the normal picture of reality, in which the idea that objects can be localised plays a crucial role. We shall return to this topic later.

Are there any other alternatives? Certainly some rather bizarre possibilities exist. The 'decision' to put the second mirror in place was made prior to the experiment with two mirrors being performed. Maybe this process somehow affected the particles used in the experiment and hence led to the observed results. Alternatively, it could in some way have affected the first mirror, so that the two mirrors 'knew about' each other and therefore behaved differently. Such things could be true, but they seem unlikely. We mention them here to emphasise how completely the results we have discussed in this chapter violate our basic concept of reality, and also because they are, in their complexity, in stark contrast to the elegant simplicity of the quantum theoretical description of these experiments. It is this description that forms the topic of the next chapter.

1.5 Summary of Chapter One

In this chapter we have discussed two separate sets of experiments associated with the passage of a particle through a potential barrier. The experiments measure different things, so the results obtained are not directly comparable and clearly cannot in themselves be contradictory. However, we have tried to justify our interest in what *actually happens* in addition to what is seen, and when we use the experiments to tell us what happens we obtain incompatible information. The first experiment tells us that particles are either transmitted or reflected by the barrier. We can therefore consider, for example, a particle that is reflected and remains always to the right of the barrier. The second experiment then tells us that in some cases the subsequent behaviour of this particle can depend on whether or not the left-hand mirror is

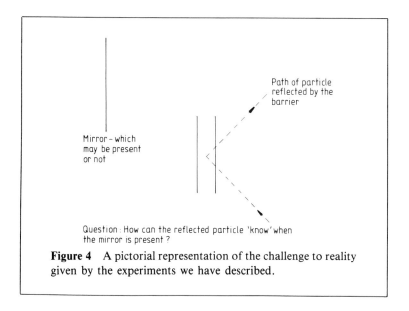

Figure 4 A pictorial representation of the challenge to reality given by the experiments we have described.

present, regardless of how far away it might be. Readers should be convinced that this is crazy—because it *is* crazy. It also happens to be true. This is the challenge to reality which is a consequence of quantum phenomena. We illustrate it, pictorially, in figure 4.

How this challenge is being met, the extent to which we can understand what is actually happening, the possible forms of reality to which quantum phenomena lead us, are the subjects that will occupy us throughout the remainder of this book.

Chapter Two

Quantum Theory

2.1 The description of a particle in quantum theory

The familiar, classical, description of a particle requires that, at all times, it exists at a particular position. Indeed, the rules of classical mechanics involve this position and allow us to calculate how it varies with time. According to quantum mechanics, however, these rules are only an approximation to the truth and are replaced by rules that do not refer explicitly to this position but, instead, predict the time variation of a quantity from which it is possible to calculate the *probability* of the particle being in a particular place. We shall indicate below the circumstances in which the classical approximation is likely to be valid.

The probability will be a positive number (any probability has to be positive) which, in general, will vary with time and with the spatial point considered. As an example, figure 5 is a graph of such a probability, and shows how it varies with the distance, denoted by x, along a straight line from some fixed point O. This graph represents a particle which is close to the point labelled P. The width of the distribution, shown in the figure as U_x, gives some idea of the uncertainty in the true position of the particle. There are precise methods of defining this uncertainty but these are not important for our purpose. Clearly a very narrow peak corresponds to accurate knowledge of the position of the particle and, conversely, a wide peak to inaccurate knowledge.

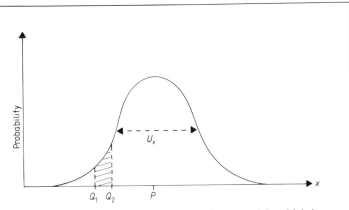

Figure 5 A typical probability graph for a particle which is
close to a point *P*. The probability of finding the particle in the
neighbourhood of any point is proportional to the height of
the curve at that point. If we measure area in units such that
the total area under the curve is one, then the probability that
the particle is in the interval from Q_1 to Q_2 is equal to the
shaded area. For a simple peak of this form the uncertainty in
position is the width of the peak, denoted here by U_x.

At this stage it might be thought that we can always use the
classical approximation, where particles have exact positions, by
working with sufficiently narrow peaks. However, if we do this we
lose something else. It turns out that the width of the peak is also
related to the uncertainty in the velocity of the particle, more
precisely the velocity in the direction of the line between the points
O and *P*, only here the relation is the opposite way round: the
narrower the peak, the larger the uncertainty. In consequence,
although there is no limit to the accuracy with which either the
position or the velocity can be fixed, the price we have to pay for
making one more definite is loss of information on the other. This
fact is known as the *Heisenberg uncertainty principle*.

Quantitatively, this principle states that the product of the
position uncertainty and the velocity uncertainty is at least as large
as a certain fixed number divided by the mass of the particle being
considered. The fixed number is, in fact, the constant $\hbar$ introduced

earlier. We can then write the uncertainty principle in the form

$$U_x U_v > \hbar/m \qquad (2.1)$$

where U_v is the uncertainty in the velocity and m is the mass of the particle.

The quantity $\hbar$ is Planck's constant. We quote again its value, this time in SI units:

$$\hbar = 1.05 \times 10^{-34} \text{ kg m}^2 \text{s}^{-1}. \qquad (2.2)$$

This is a very small number! We can now see why quantum effects are hard to see in the world of normal sized, i.e. 'macroscopic', objects. For example, we consider a particle with a mass of one gram (about the mass of a paper clip). Suppose we locate this to an accuracy such that U_x is equal to one hundredth of a centimetre (10^{-4}m). Then, according to equation (2.1), the error in velocity will be about 10^{-20} m per year. Thus we see that the uncertainty principle does not put any significant constraint on the position and velocity determinations of macroscopic objects. This is why classical mechanics is such a good approximation to the macroscopic world.

We contrast this situation with that which applies for an electron inside an atom. The uncertainty in position cannot be larger than the size of the atom, which is about 10^{-10} m. Since the electron mass is approximately 10^{-30} kg, equation (2.1) then yields a velocity uncertainty of around 10^6 m s^{-1}. This is a very large velocity, as can be seen, for example, by the fact that it corresponds to passage across the atom once every 10^{-16} s. Thus we guess, correctly, that quantum effects are very important inside atoms.

Nevertheless, readers may be objecting on the grounds that, even in the microscopic world, it is surely possible to devise experiments that will *measure* the position and velocity of a particle to a higher accuracy than that allowed by equation (2.1), and thereby demonstrate that the uncertainty principle is not correct. Such objections were made in the early days of quantum theory and were shown to be invalid. The crucial reason for this is that the measuring apparatus is also subject to the limitations of quantum theory. In consequence we find that measurement of one of the quantities to a particular accuracy automatically disturbs the other and so induces an error that satisfies equation (2.1). As a simple example of this, let us suppose that we wish to use a microscope to measure the position of a particle, as illustrated in figure 6. The microscope detects light which is reflected from the particle. This light, however, consists of photons, each of which carries momentum. Thus the velocity of the particle is continuously being altered by the light that is used to measure its position. It is not possible to calculate these changes since they depend on the directions of the photons after collision. The resulting uncertainty can be shown to be that given by the uncertainty relation. The caption to figure 6 explains this more fully. Most textbooks of quantum theory, e.g. those mentioned in the bibliography (§6.5), include a detailed analysis of this experiment and of other similar 'thought' experiments.

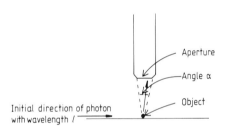

Figure 6   Showing how the uncertainty principle is operative when a microscope is used to fix a position. For an accurate measurement of position the aperture should be large, but this leads to a large uncertainty in the direction of the photon, and hence to a large uncertainty in the momentum of the object. In fact, the error in position is given by $l/\sin\alpha$ and that in momentum by $p\sin\alpha$ where p is the photon momentum, related to its wavelength by $l = 2\pi\hbar/p$ [cf equation (2.4)]. Hence the product of the errors is equal to $2\pi\hbar$, as required. Note that a crucial part of the argument here is that light is quantised, i.e. light of a given wavelength comes in quanta with a fixed momentum.

So far in this section we have taken the probability to depend upon just one variable, namely the distance x along some line. In general, of course, it will depend upon position in three-dimensional space. Nothing in the above discussion is greatly affected. The position uncertainty in any particular direction is always related by the uncertainty principle, equation (2.1), to the velocity uncertainty in the same direction.

Since we are considering one particle, which has to be somewhere, the probabilities of finding it in a particular region of space, when added over all such regions, must give unity. Because the points of space are not discrete but rather continuous, this addition is performed by an 'integral'. Most readers will probably not wish to be troubled by such technicalities so, since they are not essential for understanding the subsequent discussion, we relegate further details of this and a few other matters connected with the probability to Appendix 3. One fact will be useful for us to know. In the one-dimensional case the probability of finding the particle in any interval is equal to the area under the graph of the probability curve, bounded by that interval. This is illustrated in figure 5. Of course, in order that the total probability should be unity it is important that the area is measured in units such that the total area under the probability graph is equal to one.

To proceed we must now go beyond the probability and consider the quantity from which it is obtained. This is called the *wavefunction* and, being the basic quantity which is calculated by quantum mechanics, it will play an important part in the development of our story. What the wavefunction *means* is, as we shall see, very unclear; what it *is*, however, is really quite simple. Since it involves ideas that will be new to some readers we devote the next section to it.

2.2 The wavefunction

We consider a system of a single particle acted upon by some forces. In classical mechanics the state of the system at any time is specified by the position and velocity of the particle at that time. The subsequent motion is then uniquely determined for all future times by solution of Newton's second law of motion, which tells us that the acceleration is the force divided by the mass.

In quantum theory the state of the system is specified by a wavefunction. Instead of Newton's law we have Schrödinger's equation. This plays an analogous role because it allows the wavefunction to be uniquely determined at all times if it is known at some initial time. Thus quantum mechanics is a deterministic theory of wavefunctions, just as classical mechanics is of positions.

The wavefunction of a particle exists at all points of space. It consists of two numbers, whose values, in general, vary with the point considered. We shall find it convenient later to picture these two numbers by regarding the wavefunction as a line on a plane, like that shown in figure 7. The two numbers are then the length of the line and the angle it makes with some fixed line. We shall refer to these numbers as the magnitude and the angle of the wavefunction. Readers who wish to use the proper technical language should refer to Appendix 4.

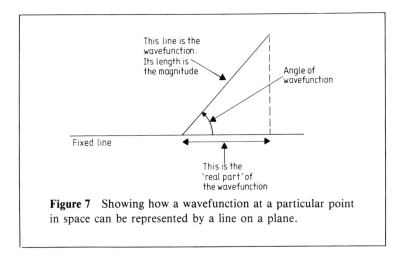

Figure 7 Showing how a wavefunction at a particular point in space can be represented by a line on a plane.

As mentioned in the previous section, the wavefunction at a given point determines the probability for the particle to be at that point. In fact, the relation between the wavefunction and the probability is very simple: the probability is proportional to the square of the magnitude of the wavefunction. It does not depend in any way on the angle of the wavefunction.

The classical notion of a particle's position is therefore related to the magnitude of the wavefunction. What about the classical velocity? Not surprisingly, this is related to the angle. In fact, the velocity is proportional to the rate at which the angle of the wavefunction varies with the point of space, i.e. with x. The reason for this is discussed in Appendix 4 (but only for readers with the necessary mathematical knowledge). Note that here we are speaking of the actual velocity, not the *uncertainty* in the velocity which, as discussed earlier, is proportional to the width of the peak in the probability.

For easier visualisation of what is happening it is useful to simplify the idea of a wavefunction by thinking about its so-called *real part*, which is the projection of the wavefunction along some fixed line, as shown in figure 7. For example, the real part of the wavefunction corresponding to the probability distribution of figure 5 might look like figure 8. The dashed line in this figure is the magnitude of the wavefunction. The rate of oscillation of the real part is proportional to the velocity of the particle.

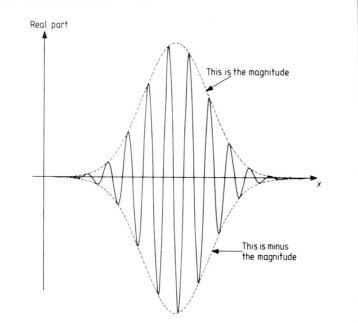

Figure 8 A typical wave packet. The broken curve indicates the magnitude of the wavefunction and the solid curve gives the 'real part'. The rate of oscillation is proportional to the average velocity of the particle.

We shall see later that it is necessary to have a method of 'adding' wavefunctions. The method we use can be understood by reference to figure 9. We wish to add the wavefunctions represented by the lines in figures 9(*a*) and (*b*). To do this we join the beginning of the first line to the end of the second; then the line joining the beginning of the second to the end of the first is the line that represents the sum of the two wavefunctions. This is illustrated in figure 9(*c*). It is not hard to show that, with this definition, it is irrelevant which line is called the first and which the second. We now notice the important fact that this definition is not the same as using ordinary addition to add the numbers associated with each wavefunction. In particular, the magnitude of the sum of two wavefunctions is not the same as the sum of the magnitudes of the wavefunctions. As an example of this, whereas, since magnitudes are always positive, the sum of two magnitudes is always greater than either, this is not necessarily the case for the magnitude of the sum, as is seen in figure 10. Note, however, that the real parts of wavefunctions do add just like ordinary numbers.

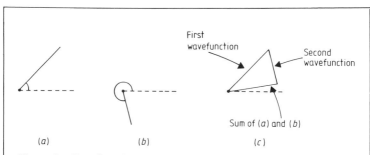

Figure 9   Showing how two wavefunctions, (*a*) and (*b*), add together to give a new wavefunction (*c*).

Readers who wish to know further mathematical details regarding wavefunctions, their addition, etc, should consult Appendix 4. Such details will not be essential for what follows.

We are now in a position to understand the quantum mechanical treatment of the two types of potential barrier experiment introduced earlier. These topics will be our concern in the next two sections.

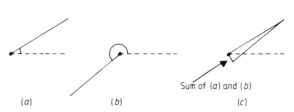

Sum of (*a*) and (*b*)

(*a*)   (*b*)   (*c*)

Figure 10   Another example of addition of two wavefunctions. We note, in particular, that the magnitude of the sum of the wavefunctions is smaller than the magnitudes of either of the two wavefunctions.

2.3   The potential barrier according to quantum mechanics

We require for this problem an initial state which corresponds as closely as possible to the classical situation, i.e. a particle on the right of the barrier and moving towards it with a velocity v. To this end we take a wavefunction with a magnitude that is peaked in the neighbourhood of the initial position and with an angular variation such that the average velocity is equal to v. There will of course be an uncertainty in both the position and the velocity, according to equation (2.1). A possible form for the square of the magnitude, which we recall is proportional to the probability, is shown in figure 11(*a*). Since we are dealing with one particle the area under this peak will be equal to one.

The Schrödinger equation now determines the subsequent behaviour of this wavefunction. We shall not discuss the method of solving the equation but merely state the results. The peak in the wavefunction moves towards the barrier with a velocity approximately v—this is very similar to the classical motion of a particle where there are no forces. There is, in addition, a small increase in the width of the peak, so the situation at a later time is shown in figure 11(*b*). When the peak reaches the barrier, where the effect of the force begins to be felt, it spreads out more rapidly and then splits into two peaks, as seen in figure 11(*c*). These two peaks then move away from the barrier in opposite directions, so a little later we have the situation shown in figure 11(*d*). Our wavefunction has separated into two peaks, one reflected and one transmitted by the barrier.

It is a consequence of the Schrödinger equation that, throughout the motion, the total area under the graph of the square of the

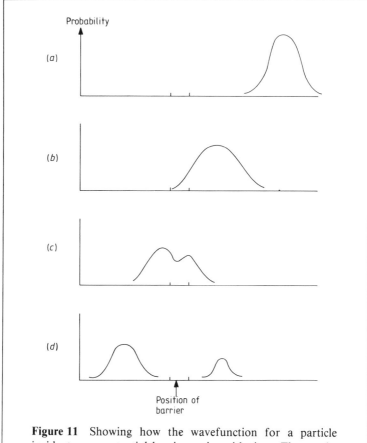

Figure 11   Showing how the wavefunction for a particle incident on a potential barrier varies with time. The graphs (*a*)–(*d*) show the square of the magnitude of the wavefunction at four successive times. The pictures correspond to a case where the probability of transmission is greater than that of reflection.

length of the wavefunction remains equal to one. In fact we know that this has to be true for consistency with the probability interpretation—the particle always has to be somewhere. The probability that it is on the right of the barrier, i.e. that it has been reflected, is given by the area under the right-hand peak, whereas the probability for transmission is given by the area under the peak on the left. Thus the calculation allows us to predict these probabilities and to compare with the results of experiments as discussed in §1.3. In all cases where calculations using the Schrödinger equation have been compared with experiment the agreement is perfect. In particular, it is worth mentioning that we obtain agreement with the classical result for a very high or very low potential barrier, namely almost 100% reflection or transmission respectively.

We must now look more closely at what our calculation for the potential barrier experiment really tells us. After collision with the barrier the wavefunction, and hence the probability, is the sum of two pieces. Here we are ignoring the fact that the two parts are in practice joined because the wavefunction is never quite zero, just very small, between them. What, then, happens when we make an observation which tells us whether the particle has been reflected? Clearly, in some sense, we 'select' one of the two peaks in the wavefunction. In other words, we might say that the wavefunction has jumped from having two peaks to having only one. This process is referred to as 'reduction of the wave packet'. What it means, whether it happens and, if so, how, are topics to which we shall return.

To close this section we emphasise that the wavefunction is determined from the initial conditions in a completely deterministic way. Knowing the initial wavefunction exactly (e.g. figure 11(*a*)), we can calculate, without any uncertainty, the wavefunction at all later times and hence the probability of transmission or reflection. The non-deterministic, probabilistic, aspects of the potential barrier experiment arise because we do not observe wavefunctions but rather particles; in particular, we can observe the position of an individual particle after it has interacted with the barrier.

2.4 Interference

We shall next consider the quantum theoretical description of the second type of barrier experiment discussed in Chapter One. In this, we recall, there were mirrors which could bring both the reflected and the transmitted particles to the same set of detectors. We begin then with the same initial state as before (figure 11(a)) and follow the wavefunction to the situation shown in figure 11(d). Here, to a good approximation, the wavefunction can be regarded as a sum of two wavefunctions, one giving the left-hand peak and the other the right-hand peak. Note that the operation of adding the two wavefunctions is rather trivial at this stage since, at any given point of space, at most one of the two wavefunctions which are added is different from zero. In the subsequent motion each of the two peaks will change independently; in fact they will move in a manner closely resembling the classical motion of a free particle. (It is irrelevant here that the area under each peak is not actually equal to one.)

Eventually, if the mirrors are present, the peaks will come together in the neighbourhood of the detectors. At this stage the addition is no longer trivial since both wavefunctions are different from zero at the same place. This means that the feature mentioned at the end of §2.2 becomes relevant, and the probability resulting from the two wavefunctions is not equal to the sum of the probabilities associated with the separate wavefunctions.

We have here an example of an extremely important phenomenon known as 'interference'. It occurs in a wide range of physical situations even where quantum effects are not relevant. As an example, we can think of two pebbles being dropped onto the surface of a still pond. Ripples will spread out from the points of impact. At some positions on the pond the 'ups' and the 'downs' from the two circular wave patterns will always come at the same time and the wave will therefore be enhanced. At others they will be 'out of phase', i.e. an 'up' from one will arrive at the same time as a 'down' from the other, in which case they will cancel each other and the water will remain still. Figure 12 illustrates this situation.

In our quantum mechanics problem the situation is rather more complicated since we are not just adding numbers, which can be positive or negative, but adding 'lines', and we recall that the result depends on the angle between the lines. On the other hand, if we think just of the real parts of the wavefunctions, then what happens is very similar to the case of water waves. The precise forms of the two wavefunctions to be added will depend on the length of the path to any particular detector (see figure 4, for example). It follows that the nature of the interference observed will depend on which detector is considered. Certainly, in general, the probability resulting from the sum of the two wavefunctions will be different from the sum of the probabilities coming from each separately.

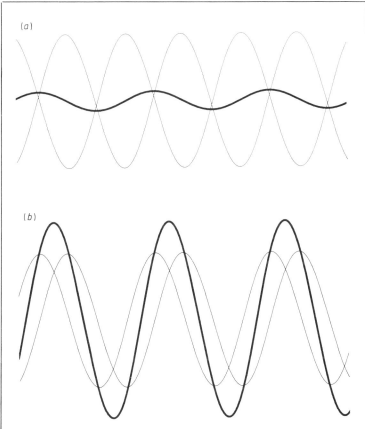

Figure 12 Illustrating the way that waves interfere. The thin lines represent the contributions of two different sources, and the heavy lines their sum, all plotted as functions of time. In (*a*) the two contributions almost exactly cancel, whereas in (*b*) they have similar phase and add to produce a larger effect.

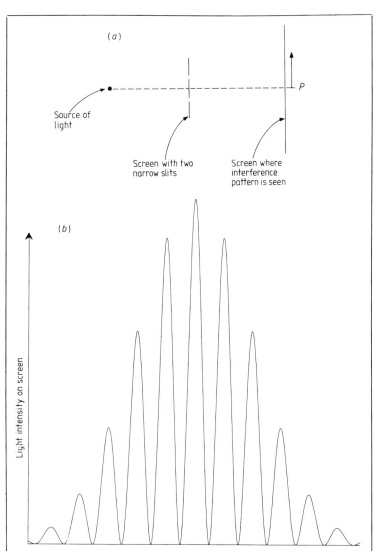

Figure 13 (*a*) An experiment which shows interference effects for electromagnetic radiation, e.g. light. The radiation from the source can reach the right-hand screen through either slit. At the point *P* the radiation will arrive in phase because the two path lengths are the same, whence there is constructive interference. At points away from *P* the path lengths are different and destructive interference is possible. A pattern of intensity like that shown in (*b*) emerges on the screen.

This is in accordance with the observations which we found so surprising in §1.4.

Detailed calculations yielding precise results are, of course, possible. Similar calculations can be done for other situations in which quantum mechanical interference occurs, and where the results can be verified by experiments. Of particular importance are experiments where electrons are scattered off crystals. Here the interference is between parts of the wavefunction scattered off different sites in the crystal. Comparison of the results with calculated predictions reveals information on the structure of the crystal.

A brief historical note is of interest here. The long-standing conflict between a corpuscular theory of light (favoured by Isaac Newton) and a wave theory was generally believed to have been settled in favour of the latter by observation of interference effects when light was passed through two slits (see figure 13). Interference implied waves. It was therefore a shock when electrons, long established as particles, were also found to show interference effects. This schizophrenic behaviour became known as 'particle–wave duality'. The same duality applies to electromagnetic radiation, of which light is an example. The 'particles' of light are called photons. In our potential barrier example, the particle nature is seen most naturally in the first set of experiments where the particle is observed either to be transmitted or reflected. The wave nature is seen in the second set, where there is evidence for interference effects.

Quantum theory successfully incorporates both features and enables us to calculate correctly all microscopic phenomena that do not involve 'relativistic' effects. A brief review of some of the successes of the theory is given in the next section, with which we conclude this chapter. The big question of what the quantum theoretical calculations actually mean is left to Chapter Three.

2.5 Other applications of quantum theory

In this section we shall outline some of the most important applications of quantum theory to various areas of physics, applications which ensured that, in spite of its problems, it rapidly gained acceptance. Nothing in the remainder of our discussion will depend on this section, so it may be omitted by readers who are in a hurry. The section is also somewhat more demanding with regard to background knowledge of physics than most.

The understanding of electricity and magnetism, besides being the prerequisite for the scientific and technological revolutions of this century, was the great culminating triumph of nineteenth century, classical, physics. By combining simple experimental laws, deduced from laboratory experiments, into a mathematically consistent scheme, Maxwell unified electric and magnetic phenomena in his equations of electromagnetism. These equations predicted the existence of electromagnetic waves capable of travelling through space with a calculable velocity. Visible light, radio waves, ultraviolet light, heat radiation, x-rays, etc, are all examples, differing only in frequency and wavelength, of such waves.

The first hint of any inadequacy within this scheme of classical physics came with the calculation of the way in which the intensity of electromagnetic radiation emitted by a 'black body' (i.e. a body that absorbs all the radiation falling upon it at a particular temperature) varies with the frequency of the radiation. The assumptions which went into the calculation were of a very general nature and were part of the accepted wisdom of classical physics; the results, however, were clearly incompatible with experiment. In particular, although there was agreement at low frequency, the calculated distribution increased continuously at high frequency rather than decreasing to zero as required.

Max Planck, in 1900, realised that one simple modification to the assumptions would put everything right, namely, that emission and absorption of radiation by a body can only occur in finite sized 'packets' of energy equal to h times the frequency. The constant of proportionality introduced here, and denoted by h, is the original Planck's constant. For various reasons it is usual now to work instead with the quantity $\hbar$, which we quoted in equation (2.2), and which is equal to h divided by 2π.

The packets of energy, introduced by Planck, are the 'quanta' which gave rise to the name quantum theory. Each such quantum is now known to be a photon, i.e. a particle of electromagnetic radiation, but such a concept was a heresy at the time of Planck's original suggestion; electromagnetic radiation (e.g. light, radio waves, etc) was known to be waves! The quantisation was therefore assumed to be simply something to do with the processes of emission and absorption.

Such a view was shown to be untenable by the observation of the photoelectric effect, in which electrons are knocked out of atoms by electromagnetic radiation. If we assume that the energy in a uniform beam of light, incident upon a plate, is distributed uniformly across the plate, then it is possible to calculate the time required for sufficient energy to fall on one atom to knock out an electron. This is normally of the order of several seconds, in contrast to the observation that the effect starts immediately. Further, the energy of the emitted electrons is, apart from a constant, proportional to the frequency of the radiation. Einstein, in 1905, showed that all the observations were in perfect agreement with the assumption that the radiation travelled as photons, each carrying the energy E appropriate to its frequency according to the relation previously used by Planck:

$$E = hf \qquad (2.3)$$

where f is the frequency.

The final confirmation of the idea of photons came from the observation, in 1922, of the Compton effect, in which radiation was seen to decrease in frequency when it was scattered by electrons. This can be explained very simply as being due to the loss of energy in the photon–electron collision, a loss that can be exactly calculated from the laws of conservation of energy and momentum.

Although quantum theory began with its application to radiation, the ideas were soon applied to particles. In 1911, de Broglie suggested that, if waves can have particle properties, then it is reasonable to expect particles to have wave properties. He introduced the relation:

$$l = h/mv \qquad (2.4)$$

between the wavelength l, the velocity v, and the mass m of a particle. The major achievements of quantum mechanics have been, following this relation, in its application to matter, in particular to the structure of atoms.

The experimental work of Rutherford, early this century, showed that an atom consists of a small, positively charged, nucleus, which contains most of the mass of the atom, surrounded by a number of negatively charged electrons which are bound to the nucleus by the attractive electric force. Each atom was therefore like a miniature solar system, with the electrons playing the role of planets, orbiting the nuclear 'sun'. Prior to the advent of quantum theory there were, however, serious problems with this picture: why did the orbiting electrons not radiate electromagnetic waves, thereby losing energy so that they would fall into the nucleus? Why were the energies available to a given atom only a set of discrete numbers, rather than a continuum as would be expected from classical mechanics?

Quantum theory provides a complete answer to these questions. All the energy levels of atoms can be calculated from the Schrödinger equation, in perfect agreement with experiment. The interactions between atoms, as observed in molecules, chemical processes and atomic scattering experiments can also be understood from this equation. As we mentioned in §1.1, quantum theory successfully brought a whole new range of phenomena into the domain of calculable physics.

The details of all this are outside the scope of this particular book, but it is worthwhile to give a simple picture of why the wave nature of the electron helps us to understand the quantum answers to the problems mentioned above with the classical picture of the atom. If we consider a wave on a string with fixed end points, then only certain wavelengths are allowed, because an integral multiple of the wavelength must fit exactly into the string. A consequence is that the string can only vibrate with a particular set of frequencies; a fact which is crucial to many musical instruments. The frequencies which occur can be altered by changing either the length or the tension of the string. In an atom the situation is similar, except that, instead of having a wave on a string with fixed end points, we have a wave on a circle (the orbit), which must join smoothly on to itself. Thus the circumference of the circle has to be an exact integral multiple of the wavelength. As we show in Appendix 5, this condition yields the energy levels of the simplest atom.

The transition from one energy level in an atom to another, by the emission of a photon, i.e. by electromagnetic radiation, is an example of an important class of very typically quantum phenomena, in which one particle spontaneously 'decays' into (say) two others. Calling the first particle A and the others B and C, we can write this as

$$A \rightarrow B + C.$$

If we start with a large number of A particles then, after a given time, some of them will have decayed. It is usual to define a 'half-life' as the time taken for half of the particles in a large initial sample to have decayed. The half-life depends on the process considered and values ranging from tiny fractions of a second to times beyond the age of the universe are known.

Even though the half-life for the decay of a certain type of particle, e.g. the A particle above, might be known, it will not be possible to say when a *particular* A particle will decay. This is random; like, for example, the choice of transmission or reflection in the potential barrier experiment. Indeed, one can think of some types of decays as being rather like a particle bouncing backwards and forwards between high potential barriers; eventually the particle passes through a barrier and decay occurs. In general, if we start with a wavefunction describing only identical A particles, then it will change into a sum of a wavefunction describing A particles,

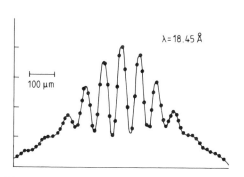

Figure 14 A comparison of the data obtained in a two-slit interference experiment, like that shown in figure 13, using neutrons. The curve is the calculated prediction of quantum theory and the experimental results are shown as dots. The agreement is perfect. (After an original figure in the paper of Zeilinger, Gaehler, Shull and Treimer *Symposium on neutron scattering* Am. Inst. Phys. 1981.)

which will have a magnitude decreasing with time, and one describing (B + C), which will have an increasing magnitude.

Finally, we mention the recent, very accurate, experiments which show that neutrons passing through a double slit, as in figure 13, interfere exactly as predicted by quantum theory. An example is shown in figure 14. These experiments were carried out in response to the recent upsurge of interest in checking carefully the validity of quantum theoretical predictions in as many circumstances as possible. We shall later mention other such tests. In all cases so far the theory is completely satisfactory.

2.6 Summary of Chapter Two

We have shown how the classical description of a particle, involving its position and velocity, is replaced by a description in terms of a wavefunction. If this wavefunction is known at some initial time, for an isolated system, then it is completely determined for all future times by the solution of the Schrödinger equation.

The relation of the wavefunction to experimental observation introduces the non-causal aspects into the problem since the wavefunction only predicts the probability of obtaining a given experimental result. For macroscopic objects the range of probabilities is effectively so small that the classical approximation is normally adequate. This, however, is certainly not true in the microscopic world, where the quantum effects are important.

We have seen in particular how quantum mechanics predicts the previously discussed results of the potential barrier experiment and have noted especially the importance of interference effects in obtaining these results. Such interference effects are also important in the many successes of quantum theory which we have discussed.

Any 'measurement' on a system described by quantum mechanics chooses one of certain possible results, i.e. it selects part of the wavefunction. This process, known as reduction of the wavefunction, will need to be considered further in the next chapter.

Chapter Three

Quantum Theory and External Reality

3.1 Review of the problem

In Chapter One we saw that there are aspects of the observed world which appear to be mutually contradictory, at least when they are interpreted in terms of our normal pictures of reality. Since such pictures are necessarily developed from experience of things that we can see and feel, that is, from the microscopic world, whereas the apparent contradictions only occur when microscopic objects are involved, we should perhaps not be too disturbed by this discovery. Our pictures of reality, the words and metaphors we use, are not necessarily appropriate for the world of the very small. The evidence of Chapter One suggests that we need new pictures.

In the second chapter we discussed quantum theory, which, developed in response to the strange phenomena seen in the microscopic world, very beautifully predicts such phenomena. It has proved to be the most successful and comprehensive theory known in physics. We are therefore naturally encouraged to expect that it might help us to understand the reality that underlies the observations of the microscopic world. How far such an expectation is realised is the topic of this and the following chapter.

We shall ask whether quantum theory has merely allowed us to calculate the results of our experiments or whether it has, in addition, answered the problems we met in the first chapter. We recall that these problems did not lie in the results themselves but arose when we asked what was actually happening. Does quantum theory tell us what happens when a particle hits a potential barrier? What does it tell us about the external reality which is present before we make our observations? Is it, indeed, even compatible with the existence of such a reality? These are some of the questions we shall try to answer.

3.2 The ensemble interpretation of quantum mechanics

As we have seen, quantum theory deals with a wavefunction, which it states is causally determined from some initial conditions. The passage from this wavefunction to experimental observation uses the assumption that the wavefunction gives probabilities for measurements to yield particular values. In order to test the predictions of the theory it is necessary to prepare a large number of *identical* systems and perform the same measurement on each. We recall that we used this procedure to define the probabilities of transmission and reflection in §1.3. Of course, the word identical now must refer to the wavefunction, i.e. 'identical systems' are defined to be systems with the same initial wavefunction (and therefore the same wavefunction for all future times).

The large number of identical systems is referred to as an ensemble. For any such ensemble the predictions of quantum theory are precise and deterministic. For example, quantum theory tells us what percentage of a given (large) number of particles will pass through a potential barrier. What it cannot tell us, of course, is whether any particular one of the particles will pass through.

Some writers on this topic have therefore adopted the view that quantum theory is a *theory of ensembles* and as such tells us nothing about individual systems. This is a perfectly reasonable view and it may be the correct one to take. There are then no further difficulties in the 'interpretation of quantum theory', and the subject does not cause any philosophical problems. We must not, however, go on from this to claim that we have *solved* the problems met in the first chapter. We have merely *ignored* them. We do not only have experimental results for ensembles. Individual systems exist and the problems arise when we observe them. It is possible to argue that quantum theory says nothing about such individual systems but, even if this is true, the problems do not go away.

We shall, in this chapter, adopt a more positive view and continue to hope that the theory which predicts our results might also help to explain them.

3.3   The wavefunction as a measure of our knowledge

In §2.2 we tended to regard the wavefunction as describing a particular system (in fact, just a single particle). Suppose, however, we take the view that the wavefunction instead describes *our knowledge of the system*. By implication the system might then be thought of to have other properties of which, at the time, we are ignorant. For example, the particle may actually be at a specific position, whereas we know only that it has a certain probability of being in some region. This, at first sight, appears to be a very reasonable view. It is indeed the situation that occurs whenever probability aspects arise in non-quantum situations.

To have a trivial example of this, let us suppose that I am in a room with 1000 other people. Assume also that I know the 1000 is made up of 498 French men, 2 French girls, 200 Norwegian men and 300 Norwegian girls. With this information I would know that the probability of the person immediately behind me being French was one in two. Now suppose that I looked at the person behind me and saw that she was female. The probability of the person being French would immediately change to one in fifty.

If the situation in quantum theory is of a similar nature then the issue of the reduction of the wavefunction, raised in §2.3, immediately goes away. When the wavefunction is just an expression of our knowledge of the truth, then it is not surprising, and is even expected, that is should suddenly change to something else when a measurement is made. A measurement has simply changed our knowledge (this of course is normally the purpose of making measurements).

Superficially attractive though this view of the wavefunction may be, it is in one very important respect inadequate. It cannot explain the phenomenon of interference. We remind ourselves here that there is abundant experimental evidence for interference effects and, contrary to what appears to happen in some discussions of the interpretation problems of quantum theory, they cannot be ignored. Wavefunctions which merely represent our knowledge of a system cannot interfere. We can see this immediately in the case of the potential barrier experiment. There we require that 'something' follows both routes to the detector. That 'something' cannot be our knowledge, which, if it is anywhere, is in our brain. If the particle really has followed one route then we are back with the problem as to how its motion can be influenced by the presence of the other mirror. It is not an answer to this to say that *we* know about the other mirror; the behaviour of the particles surely cannot depend upon the information contained in the brains of particular individuals.

We can therefore be sure that, if interference actually occurs, this interpretation of the wavefunction must be wrong. However, the form of the qualification used here is important. What we *know* is that the results of our observation can be predicted from the calculation of the interference effect. It *looks* as though interference is actually happening but it is possible that this is not so, but that, instead, the calculation just 'happens' to give the right answer. A simple analogy might help here. An umpire at a cricket match counts the number of balls that have been bowled by placing pebbles in his pocket, one for each ball. When six pebbles are in the pocket he calls 'over' and play changes ends. Now the reason for this change is not directly anything to do with pebbles in the umpire's pocket, it is because six balls have been bowled and the rules say that play changes ends every six balls. The pebbles can be used by the umpire to make the calculation because of the rules of arithmetic which ensure that the right answer will be obtained. It could be that a similar thing is happening with the interference calculation; it gives the right answer but the real reason for the experimental facts lies elsewhere.

Where? Clearly we must look at the hidden information—at the properties not contained in our knowledge of the system, and therefore not in the wavefunction. We are then in the domain of hidden variable theories which we discuss in detail in Chapter Five. However, to complete this section we should look ahead and note that such theories do not in fact eliminate the need for an interfering wavefunction. Indeed, it is inconceivable that any theory could successfully reproduce all the correct effects of interference unless the interference actually happens. Thus, although it was important to mention the reservations of the previous paragraph, I believe they can now be forgotten.

3.4 The wavefunction as part of external reality

We now want to consider the possibility that the wavefunction should be treated rather more seriously than in the preceding two sections, so that we can use it to tell us something about the external reality. We shall try to regard the wavefunction not as just a description of a statistical ensemble, as in §3.1, or as a catalogue of our information about a system, as in §3.2, but as something that really exists, something that is, indeed, part of the external reality which we observe.

There are at least three good reasons why we should want to consider this assumption. First, since the classical picture of a single particle, always having a precise position and following a specific path, is not compatible with the observations described in §1.4, we do not have any other object available for our representation of reality. Secondly, the evidence that wavefunctions can interfere strongly suggests that they are real, e.g. just like ripples on the surface of a pond. In order to understand the third reason we need to know about certain symmetry properties that have to be imposed on wavefunctions describing more than one particle. If we have two *identical* particles, e.g. two electrons, then in classical mechanics we could distinguish them, for example, by their positions. In quantum theory, on the other hand, they are described by a wavefunction which tells us the probability of finding an electron at one place and an electron at another place; in no way are the two electrons distinguished. This means that the wavefunction must be *symmetrical* in the two electrons, i.e. it must not change if we interchange them. Actually, the truth is a little different from this because in some particular cases the wavefunction has to change its sign. Such a change, however, does not alter any of the physics, which is determined by the square of the magnitude of the wavefunction. A more detailed discussion of this is given in Appendix 4. Here we merely note that the symmetry properties give rise to important, testable, predictions, which have been verified and which would be very hard to understand without the assumption that wavefunctions have a real existence.

Our tentative picture of the potential barrier experiment is therefore that of a wavefunction which has a value that varies with the point of space being considered. We are familiar with quantities of this type, e.g. the temperature of the air at different points of a room, or the number of flies per unit volume in a field of cattle. Actually the wavefunction is a little different since, as we recall, it is a line or, alternatively, two numbers at each point of space. This fact, however, does not affect the present discussion, so we shall continue to refer simply to the value of the wavefunction.

As is illustrated for example in figure 11, the wavefunction is in general not constant but changes with time. Again this is a concept with which we are familiar; the temperatures at various points in a room, for example, will similarly change with time, e.g. when the heating has been switched off. We therefore have a simple picture of reality, with the wavefunction describing something that actually happens.

There are, however, two difficulties associated with this picture. The first of these is due to the fact that the world does not consist of just one particle. We remember that the wavefunction we have used so far was specifically designed to treat only one particle. How do we generalise this to accommodate additional particles?

Consider a world of two particles, which we shall call A and B. As a first guess we might try having a wavefunction for particle A and a separate and independent one for particle B. Then the probability of finding A at some point would not depend on the position of B. This is reasonable for particles that are genuinely independent, i.e. not interacting. It is, however, quite unreasonable, and is indeed *false*, for particles that are interacting. In this case the wavefunction must depend on *two* positions. It will then tell us the probability for finding particle A at one position and particle B at the other. (Some further details are given in Appendix 4.) One can express this by saying that the wavefunction does not exist in the usual space of three dimensions but in a space of two-times-three dimensions. It is no longer true to say that at a particular point of space the wavefunction has a particular value. Rather we have to say that, associated with every two points of space (or, if we prefer to express it this way, with every point of a six-dimensional space) there is a particular value for the wavefunction.

Of course, we cannot stop at two particles and must go on to include 3, 4, etc, with the wavefunction depending on the corresponding number of points, 9, 12, etc, in space. At this stage the wavefunction starts to look more like a mathematical device than something that is part of the real world. Certainly it is not now of the form of the familiar quantities mentioned earlier. These are *local*, i.e. at a single point of space there is a number which is the temperature. The wavefunction, on the contrary, is *non-local*; in order to establish its value we need to give many positions in space. We shall find this non-locality occurring again in our discussion.

It should be noted here that the two-particle wavefunction is not, in general, simply a product of two one-particle wavefunctions. To understand this distinction we recall that the square of the magnitude of the wavefunction gives the probability of finding a particle at each of the two points. If the particles are quite independent, and not in any way correlated in position, then the probability of finding a particle at a point *P* will not depend on the position of the other. In such a case the wavefunction will be a simple product of two wavefunctions, each depending upon one position. In most real situations, however, particles interact and therefore their positions are correlated. The wavefunction is then not of the product type but is, rather, one function with an explicit dependence upon two positions. Again we refer to Appendix 4 for further details.

The second difficulty that arises when we regard the wavefunction as part of reality is one to which we have already referred, the process of reduction of the wavefunction. As we saw in §2.3, the wavefunction changes when a measurement is made. This change appears to be sudden and discontinuous. It is also very non-local in the sense that measurements at one point of space can change the wavefunction at other points, regardless of how far away these might be. The measurement by means of a detector on the right-hand side of the potential barrier provides a good example of this. If this flashes it means that the particle has been reflected, so the piece of the wavefunction on the left (e.g. in figure 11) immediately becomes zero. This, at least, *appears* to be what is happening.

Whenever a measurement is made on a system described by a

wavefunction, then one of the possible values consistent with the probability distribution is obtained. The measurement somehow selects part of the wavefunction. We cannot be content, however, with merely postulating that this happens. We must ask *how* it happens. In particular, we have claimed that quantum mechanics is a universal theory and applies to everything. It should therefore apply to the apparatus which we use to make a 'measurement', and should, therefore, contain the answer to our question—that is, quantum mechanics should be able to explain how the wavefunction reduces. In fact, however, it says very clearly that the wavefunction cannot reduce! Such a startling fact deserves another section.

3.5   Measurement in quantum theory

As we have seen, it is not normally correct to say that a particle, described by quantum theory, is at a particular position. Rather, the particle has a wavefunction which tells us the probability of finding it at any given position when a measurement of position is made. Similarly, the wavefunction tells us the probability of obtaining a given value for the velocity if we make a velocity measurement. Thus measurements play a more positive role in quantum theory than in classical physics because they are not merely observations of something already present, they actually help to produce it.

A measuring instrument can be defined as something that enables us to make a measurement of the above type. That such instruments exist follows from the fact that we do actually make such measurements. We would, of course, like to believe that the apparatuses can be described by physics, i.e. that they too satisfy the rules of quantum theory. It is, however, very easy to show that this is impossible. An instrument that is able to make a measurement, in the above sense, cannot be completely described by quantum theory.

To illustrate this fact we shall consider again the potential barrier experiment with the two detectors in position. Recall that the left-hand detector records the passage of a transmitted particle and the right-hand detector the passage of a reflected particle. We suppose that each detector is a simple quantum mechanical system that can exist in one of two states OFF and ON, and that the transition between these is caused by the passage of a particle through the detector.

The complete experiment is now described by a wavefunction which contains information about both detectors as well as about the particle. Thus, for example, it would tell us the probability of

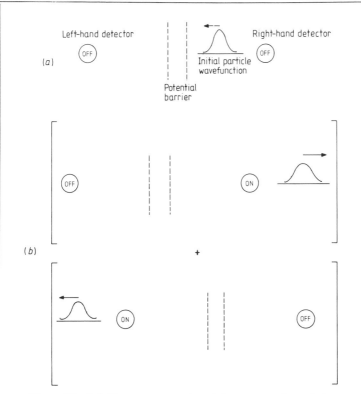

Figure 15 (*a*) Here we see a pictorial representation of the initial state in our experiment. Both detectors are in the OFF position. (*b*) This shows the final form of the wavefunction corresponding to (*a*). The measurement is a good one, in that the detectors are correctly correlated with transmission or reflection of the particle. However, the wavefunction still contains both possibilities—as in figure 11(*d*). No selection has been made.

finding the particle at a given position, with one detector in the OFF position and the other in the ON position, etc. We know the initial form of this wavefunction; it describes the particle as being incident from the right and both detectors being in the OFF position. A pictorial representation of this is given in figure 15(*a*).

The system now evolves with time according to the Schrödinger equation. This equation is more complicated than before because it must include the interaction between the detectors and the particle. We are assuming that this interaction only occurs when the particle is in the neighbourhood of a detector, and that its effect is to change the detector from OFF to ON as the particle passes through. The precise details here are not important. We can then go to a later time when the particle will certainly have passed through one detector, i.e. the two parts of the wavefunction shown in figure 11 have passed beyond the positions of the detectors. The wavefunction will now be the sum of two pieces (compare the discussion given earlier). The first piece describes a peak travelling to the right, with the right-hand detector ON and the left-hand detector OFF. The second describes a peak travelling to the left, with the right-hand detector OFF and the left-hand detector ON. Figure 15(*b*) gives a picture of this wavefunction.

We notice, first, that our measuring instruments are doing their job properly in the classical sense, that is they correctly correlate the ON/OFF positions of the detectors with the reflection/transmission of the particle. However they have not *selected* one *or* the other; the wavefunction still contains both possibilities and has not been reduced. Thus we have not succeeded in making a proper measurement in the quantum theoretical sense as we described it at the beginning of this section. Such a measurement would have left us with a final state expressible as *either* the left-hand detector ON and the right-hand detector OFF, or the other way round (with a certain probability) and not as the sum of both. Pictorially, the wavefunction would have had the form of figure 16 rather than figure 15(*b*). Readers who wish to see the difference expressed in terms of mathematical symbols should consult §4.5.

It is important now to realise that the difference between these two forms of wavefunction is not just 'words' (or even, in §4.5, 'symbols'). They are different. The difference can be seen from the fact that, at least in principle (see next section), the two parts of the sum can be brought together and made to interfere. Such interference is not possible if the wavefunction has become just one of the two pieces.

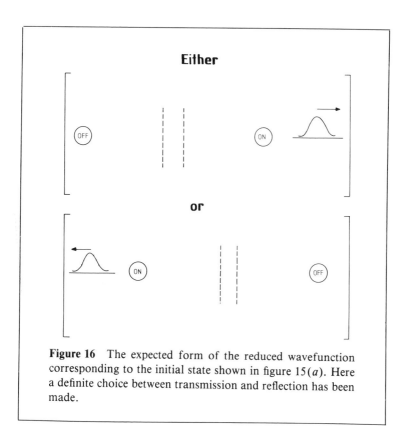

Figure 16 The expected form of the reduced wavefunction corresponding to the initial state shown in figure 15(*a*). Here a definite choice between transmission and reflection has been made.

The result we have obtained, that quantum theory does not allow the reduction of the wavefunction, is extremely important. We have obtained it in a very specialised and idealised situation, but in fact it is a completely general result. A wavefunction that can be expressed as a sum of several terms, like that of figure 15(*b*), is call-ed a *pure state*. One that is expressed as a selection of alternative possibilities, like that of figure 16, is called a *mixed state*. From the laws of quantum theory it is possible to prove that a pure state can-not change into a mixed state. Thus the wavefunction can never be reduced. An easy way to understand why this is so is to recall that wavefunctions change with time in a deterministic way, as long as they are described by quantum theory, hence they can never give the probabilistic aspects associated with measurements.

Note that we cannot solve our problem by saying, in the potential barrier example, that all we need to do is look at the detectors to see whether they are ON or OFF. This is equivalent to saying that we *measure* the state of the detectors. We then have to repeat the process and describe the new measuring apparatus, e.g. our eyes, by quantum theory. The resulting wavefunction now contains information describing this additional apparatus. It will remain a pure state.

Quantum theory, therefore, when applied to individual systems, contains an internal contradiction. It cannot describe instruments suitable for making measurements.

Faced with this situation, and bearing in mind the enormous success of quantum theory, it is natural that we should seek to modify it in such a way as to leave its successful predictions unchanged and yet to allow wavefunction reduction in appropriate circumstances. Attempts along these lines will be described in §3.7, and we shall see that there are formidable problems.

Are there any alternatives? Well, if quantum theory says that wavefunctions do not reduce we should look again at why we need them to reduce in the first place. *Why* must measurements choose? How do we know that a detector will tell us that a particle either passed through or not? The obvious answer is that we are conscious of seeing only one result. Our conscious minds do not contain both parts of the wavefunction. Maybe, then, in order to understand what is happening, we need to examine this answer more closely and to consider the concept of consciousness. This we do in Chapter Four.

3.6   Interference and macroscopic objects

We have stressed that a wavefunction which contains a sum of several terms (a pure state) is genuinely different from a wavefunction which is either one term or another (a mixed state). The reason for the difference is that, in principle, it is possible to arrange that two terms in a sum interfere when a particular probability is calculated, and this interference can be observed. Indeed, as we have seen, e.g. in §2.5, there are many experiments where this interference has been measured and found to agree perfectly with the predictions of quantum theory. However, in practice, in many situations, and in all cases where macroscopic apparatus is involved, it is not possible to design a suitable experiment to observe the interference, so the two wavefunctions are effectively indistinguishable.

To understand why this is so, let us suppose we want to check that the wavefunction for the barrier-plus-two-detectors experiment really does contain the sum of two pieces, e.g. as in figure 15(*b*). To this end we would like to arrange that the two pieces are allowed to interfere. Thus, in effect, we need to do both the potential barrier experiments of Chapter One in the same experiment. However, even when the waves corresponding to the reflected and transmitted particles are brought together by mirrors they will not interfere because, unlike the situation in §1.4, they now contain different states of the detectors (ON/OFF or OFF/ON respectively). In order to have interference it is necessary that the detectors be brought to the same state. At first sight this might appear to be easy; they can be switched to the OFF position, say. However, in order to have interference the states must be *identical*, and for macroscopic objects that is not possible. To reverse *exactly* the process whereby one of the detectors was switched to ON is, by many orders of magnitude, outside the range of any conceivable experimental technique; there is, for example, no conceivable mechanical interaction between macroscopic objects that does not remove a few atoms, slightly change the temperature of the object, alter its shape, etc. This is the reason why interference between macroscopic objects cannot be experimentally verified.

For a proper treatment of this topic we would need to use the mathematical formalism of quantum mechanics. Some of the ideas are discussed further in Appendix 6. Here we shall be content with the above rather sketchy outline of the argument. The key to it, to which we shall return, is the inherently irreversible nature of macroscopic changes.

It is clear that there is a continuum of scales ranging from the micro- to the macroscopic, so we naturally ask how far towards the latter we can go with interference experiments. At present, it seems as though the answer is not very far: all experiments so far performed deal with 'elementary' particles, or, more precisely, with systems that, for the purpose of the experiment considered, can be regarded as having very few degrees of freedom. Some ideas for doing interference experiments with larger systems are being explored at the present time (see §6.3).

3.7   Can quantum mechanics be changed so that it will reduce wavefunctions?

In the quantum theoretical decription of the potential barrier experiment, which we discussed in §§2.3 and 2.4, the wavefunction split into two pieces, one travelling to the left and one to the right. This behaviour was good because both pieces were needed to explain the interference effects. However, it is possible to modify the Schrödinger equation so that, after a certain time, the form of the wavefunction changes: one of the peaks grows and the other falls to zero. If, for example, it is the right-hand peak that remains, then the equation will have predicted that the particle is reflected. In this way, reduction of the wavefunction becomes a consequence of the modified equation. To obtain the probabilistic element which is vital for agreement with observation it is necessary that the modification contains some randomly chosen contribution. Then it is possible to arrange things so that either one of the peaks remains, with a probability proportional to its original area. In this way we obtain complete agreement with observation, and we have automatic reduction of the wavefunction.

Actually, what we have described here for one particular example can be done in general. Suitable additional terms can be added to the Schrödinger equation, so that the wavefunction automatically reduces to the form associated with a particular value for some measured quantity, always with the correct probability distribution. These extra terms must contain a random input. There is also a free constant which can be used to fix the overall magnitude of the new effects; this determines how long it takes for the reduction to occur. Some further details of the very pretty mathematics involved are given in Appendix 7.

At first sight all this appears to be just what we require for a theory of wavefunction reduction. On closer examination, however, it is clearly seen to be very unsatisfactory. The first reason for this concerns the time scale which is required for the reduction to occur. As noted above, this can be adjusted to any desired value by suitable choice of the magnitude of the extra terms in the equation. However, no choice can satisfy the experimental constraints, because these are mutually contradictory. On the one hand, it is sometimes observed that reduction takes place very rapidly, whereas, on the other hand, the observation of interference effects from radio waves that have travelled distances of the order of the size of the galaxy requires that the reduction time must be very long. No time scale for automatic reduction of the wavefunction is compatible with all observations.

The second reason why these ideas are unsatisfactory is that the wavefunction has to reduce to a form appropriate for any type of measurement. Hence the particular terms that have to be put into the Schrödinger equation depend upon what is going to be measured. In our example we have always thought in terms of position measurements, but we could instead decide to measure velocities. This would require a very different type of wavefunction reduction.

It is worth introducing here another type of experiment, totally different from anything we have met before, which illustrates this last point very well and which will also be of use later. Many particles have a 'spin', which always has a constant magnitude. For example, we shall consider electrons, where the magnitude of the spin, measured in suitable units, is always 1/2. (Appendix 8 gives some further details.) The only variable associated with the spin is its direction. (It is convenient to think of this as the direction of the axis of a spinning top.) In order to ascertain this direction we measure the spin along any line in space. It is a consequence of quantum theory that, in such a measurement, we will always find one of two values, $+1/2$, corresponding to the spin being along the chosen line, and $-1/2$, corresponding to its being in the opposite direction (see figure 17). Thus, when we make a measurement, the wavefunction will reduce to the form corresponding to plus or minus 1/2 along the line chosen. As we have stated above, it is possible for this wavefunction reduction to happen automatically if quantum theory is suitably modified. However, the final form of the reduced wavefunction, and therefore the modification required, will depend upon which particular line in space is chosen for the measurement. There cannot be one equation which describes the future evolution of the electron wavefunction, regardless of what we choose to measure.

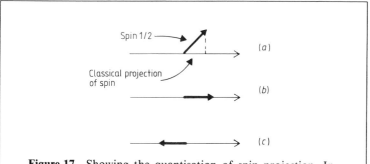

Figure 17   Showing the quantisation of spin projection. In (*a*) we see that the spin can lie in any direction according to classical physics, and the projection of the spin is given by simple geometry. The situation in quantum theory, on the contrary, is that the projection can only take the two values shown in (*b*) or (*c*).

It is clear that both these objections to the type of theory involving automatic reduction of the wavefunction can be met if the modifications to the standard quantum theory 'know about' what is to be measured and when. In other words, the new Schrödinger equation must depend upon the form of all the apparatus involved, including the measuring instruments and, for example, whatever (or whoever) decides on the direction for a spin determination. The work that we have outlined above suggests that theories of this type might be possible, but much work remains to be done and there is a danger that what emerges will look more like an arbitrary prescription to obtain the results than like a proper theory. Certainly it is hard to see how it can look at all natural.

There are three other points which might be relevant to this section and which certainly should be mentioned. First, all real measuring instruments are *macroscopic*. To appreciate how different such an object is from a single electron, say, we should realise that an object with a mass of one kilogram contains about 10^{27} particles. It is, therefore, not hard to imagine that effects which are utterly negligible for single particles might build up to something important for macroscopic objects. Two particular ways in which the mass of an object might appear in the formulae for reduction are suggested at the end of Appendix 7.

Secondly, as we have seen in the previous section, all tests of interference effects refer to particles. It is just not possible to test whether they would also occur for macroscopic objects where a very large number of degrees of freedom are involved. The difference between whether they really do occur, as predicted by quantum theory, or whether they do not, has no obvious measurable consequences. This is unfortunate, because the question has enormous relevance to the issues we are discussing.

Finally, if it is true that really new effects arise for large, complex, systems, then we should ask whether there are other manifestations of these. Is it even possible that one such effect could be consciousness, which might also be expected to occur only for large systems? Maybe, somewhere here, there is a link between this section and the subject of our next chapter.

3.8   Summary of Chapter Three

We have asked what, if anything, quantum theory tells us about external reality. The answer depends on how we regard the wavefunction. If the wavefunction is only relevant to ensembles, or if it is simply a statement about our knowlege of the individual system it is used to describe, then the theory does not help us in our search for the true nature of reality.

On the other hand, if the wavefunction of an individual system is something that is part of the external reality, we meet the problem of how such a wavefunction can be reduced by a measurement. We have discovered the important result that this cannot happen if the measuring instruments satisfy the laws of quantum theory. Thus, somewhere within quantum theory there appears to be an inconsistency; the theory tells us about the results of measurements but instruments that are capable of making such measurements cannot obey its laws.

Faced with this situation we have considered the possibility that the laws might be suitably modified. Such modifications might not seriously affect the behaviour of microscopic systems, but could perhaps cause the necessary reduction of the wavefunction when account is taken of the macroscopic nature of real measuring instruments. However, the available models of this type, although interesting, are still at a tentative stage and much further development is needed before they can be regarded as convincing

Chapter Four

Consciousness

4.1 The relevance of conscious observers

In §3.5 we proved that an instrument governed by the laws of quantum theory is not capable of making a proper measurement, that is, it cannot cause the wavefunction of a system to change to a state corresponding to a particular value of the quantity to be measured. As an example we saw that, in the potential barrier experiment, even after the attempted measurement of transmission or reflection, the wavefunction still contained pieces corresponding to both possibilities.

We must not, of course, conclude from this that true measurements are impossible. We know that they occur. We can observe which of the two detectors flashes and hence deduce whether or not a particle has passed through the barrier. Our brain certainly does not permit both possibilities. Thus, although a simple, microscopic, instrument, obeying the laws of quantum theory, does not reduce wavefunctions, they are certainly reduced by the time the information reaches our brain.

What, then, is responsible for the reduction and what are the characteristics of 'instruments' that are able to cause it? We do not know the answers to these questions. It could be that, with increasing complexity and size, correction terms in the equations of quantum mechanics become more significant, so that any macroscopic apparatus can do the reduction. On the other hand, it could be that something totally new is required and that some things possess it whereas others do not. In either case it is an obvious question to ask whether there are other features of wavefunction-reducing systems that distinguish them from simpler systems. One obvious possibility that arises here is to go to the extreme end of the chain of observation and consider the possibility that the reduction does not occur until we *know* that it must, i.e. that it only occurs when *conscious* observers are involved.

Such a wild suggestion tends to horrify the austere minds of most physicists. We fear that it takes our subject, beloved for its high standards of objectivity, rigour, precision and experimental support, into a realm where nothing can be properly defined, where feelings and personality replace detached measurement, even, perhaps, to put it on a par with astrology and the reading of tea leaves! From another point of view, however, it should perhaps be seen as an exciting new development. Maybe it allows the possibility that the enormously successful methodology of physics might enter a totally new field of investigation. This would be a revolution that would, in its significance, dwarf those to which we referred in §1.1. Although it is probably fair to say that such a revolution is unlikely, we should, before dismissing it entirely, remember that J C Maxwell, the creator of the theory of electromagnetism and undoubtedly one of the greatest physicists of all time, once expressed the view that the study of atoms would be forever outside the scope of physics! Such a precedent will guard us from making similar rash statements about consciousness.

If we are to consider seriously the relevance of consciousness in the collapse of wavefunctions we must ask, and at least try to answer, the question of what it is. To this topic we turn in our next section.

4.2   What is a conscious observer?

Consciousness or 'awareness' is something we, as people, possess. We talk about it; we have a vague understanding of what it is; through it we experience many emotions, happiness, sorrow, jealousy, love, etc; we develop concepts like free will and purpose which really have no meaning without it; we can even refer to its absence, e.g. to 'unconscious' decisions, etc; but we do not have any way of defining it. It cannot be expressed in terms of other things or even be likened to other properties. It is something unique and totally different from anything else.

To discuss it, we should therefore begin with what we *know*. Or, rather, *I* should begin with what *I* know.

I am conscious. This fact I can express in an alternative way by saying that I have a conscious mind, or that my consciousness exists. However expressed, and regardless of the fact that we do not really know precisely what the statements mean, the truth they convey cannot be denied. Even if I wish to deny reality to everything else, which, as we saw in §1.2, is logically possible even if rather pointless, I cannot deny the reality of my own thoughts.

As a natural extrapolation of my experience, it is reasonable that I should assume that you, my readers, are conscious, and then to extend this to all other people. Already, however, there are those who would question this. A Princeton University psychologist, Janes, has written a book in which he claims that consciousness is a comparatively recent feature of the human race. (The book is called *The Origin of Consciousness in the Breakdown of the Bicameral Mind* [Harmondsworth: Penguin 1980]. Though I am fairly convinced that I do not believe the claim, it is expertly, and interestingly, argued.)

Having agreed that we possess consciousness, do we know what it is? It is a private 'space' in which each of us rules alone, and into which we can introduce whatever we desire of real things, i.e. those we believe exist elsewhere, or abstract things which are purely our creation. But does such a vague description allow us to say where else it might exist, who, or what, might possess it? People? Yes, by extension of ourselves. But dogs? worms? amoebae?

I hope my readers will allow me a personal note here. I remember, as a schoolboy, sitting by a riverside listening to a skylark. I think I should have been revising for examinations but, instead, I doodled some verses of poetry. Though I can just about remember them, they are inferior to the precedent I was following, so I will not expose them to public view. I mention them because in the first verse I asked whether the bird was singing because it was happy, in the second I wondered whether it was instead singing in response to feelings of sadness, but then I wondered whether it was neither, whether in fact the skylark was capable of feeling either happy or sad or whether it possessed any awareness of anything:

'Is't only nature's law that makes thee want to sing?' I was asking myself, perhaps for the first time, the question about consciousness that I have asked many times since. I still do not know the answer, and I have no idea how to go about finding it.

The nature of the problem here can be demonstrated by the following thought experiment (which could, with a little expense, even be a real experiment). Suppose we devised a series of *tests for consciousness*. A conscious being would, for example, be expected to show pleasure in some suitable way when it was praised, it would back away from any object that hit it, or otherwise showed threatening behaviour, it would seek 'food', i.e. whatever it required to sustain its activity, when needed, and would express the need urgently if the search proved unsuccessful. The list could easily be extended. Whatever property of this type we included, however, it is easy to see that we could design a computer–robot to make all the correct responses. Such a machine would pass our tests for consciousness. I believe, though I am not sure why, that it would nevertheless not be conscious. Somehow 'physical' systems, even when designed to have the attributes of consciousness, do not seem to us to *be* conscious. Thus, although it is easy to *simulate* the effects of consciousness, we should avoid making the mistake of believing that in so doing we have *created* consciousness.

Conversely, it would be possible, by careful analysis of what happens in the human brain, to correlate the various feelings like joy, sadness, anger, etc, which we associate with consciousness, with particular chemical or physical processes in the body, the release of various hormones, and such like. But surely joy is not a chemical compound, or a particular pattern of electrical currents. Or is it? Or is it just *caused* by particular physical processes occurring in the right place? Alternatively, is the truth nearer to the statement that the thoughts of the conscious mind *cause* the appropriate currents to flow? Are the emotions, or their material effects, primary?

Certainly conscious thoughts appear to have physical effects. I have just made a conscious decision to write these particular words in my word processor. The fact that you are reading them is evidence that my thoughts had real effect in the physical world. In one sense, of course, this could be an illusion (whatever that might mean in this context). The process of my writing these words could be entirely a consequence of all the particles that make up my hand, brain, etc, moving inexorably according to the laws of motion. Somewhere along the series of events in my body that leads to the typing, particular things happen that make me think I am 'deciding' what to write. But what is cause, and what merely effect?

The problems we are discussing here are more than simply a question of the language that we use to describe things. There are of course 'language' issues. For example, we could describe a pocket calculator as a machine that 'allows particular currents to flow', or, alternatively, as a machine that 'does arithmetic'. These are different sets of words describing the same thing. Our concern is more with the question of whether the calcualator *knows* that it is doing anything at all. That *real* issues are involved can be seen from the fact that our behaviour to some extent depends on how we answer these questions. Part of the reason for the concern we sometimes (too rarely) feel for people, animals, ... is that we believe these creatures are conscious.

It is outside the range of this particular book, and beyond the ability of its author, to take this discussion any further. Much has been written on the subject. In this sense it is rather like the interpretation problem of quantum theory. I have the impression that the two topics are similar in another sense—very little is understood of either!

We close this section by offering three possible 'answers' to the question of what makes an object conscious.

(*a*) Consciousness might be caused by 'complexity'. A system that is sufficiently large, with enough degrees of freedom and enough interconnections between its various parts, might be conscious. We would then expect that there would be degrees of consciousness; it would first occur in systems less complex than the human brain, and presumably would exist in a more developed form in other, as yet unknown (?) creatures. Within this sort of framework we could allow the idea that it is the correction terms to quantum theory, introduced in §3.6, that are the essential physical ingredient of consciousness.

(*b*) Something 'new' and outside physics (at least as understood at present) could be required. Then some objects could have it and some not. We could ask where in the human body it exists. Does it have a location? Does it have to have a location? If not, what makes my consciousness 'mine'?

(*c*) Perhaps the truth is that everything is conscious. Is it purely our arrogance that denies consciousness to a stone or an electron? Maybe in our potential barrier experiment particles really do 'choose' whether or not to pass through, having first assessed the situation and then consciously decided on which side they would prefer to be.

The truth might well be different from any of these and may not be properly expressible in the terms we are using. Somehow the human mind, though it is capable of understanding much that is outside itself, meets only mystery when it looks within.

4.3   Does wavefunction reduction require conscious observers?

... does the human body deviate from the laws of physics, as gleaned from the study of inanimate nature? The traditional answer to this question is 'No': the body influences the mind but the mind does not influence the body. Yet at least two reasons can be given to support the opposite thesis.†

Here we shall examine more closely the possibility that external reality consists of a wavefunction and that this wavefunction only reduces when an observation is made by a *conscious* observer. It is the existence of consciousness that introduces the probabilistic aspects into the quantum world.

One reason for at least considering such a suggestion is that wavefunction reduction (which is an inexplicable phenomenon associated with quantum theory) and decision making (which is an inexplicable phenomenon associated with the conscious mind) share the common feature of producing an increase of information; in both something previously 'unknown' becomes 'known'. More specific motivation arise, as we have seen, from the following facts. 'Simple' systems apparently do not reduce wavefunctions; they are also not conscious (we ignore here the possibility (*c*) of the previous section). At some stage of 'complexity', and for reasons that may be solely due to complexity or may involve something totally new,

†E P Wigner, *The Scientist Speculates* ed I J Good [London: Heinemann 1962]

there are systems that do reduce wavefunctions, and there are systems that are conscious. In both cases our certainty here is due to our own experience. We experience effects which appear to correspond to reduced wavefunctions and we are certainly conscious. It is therefore reasonable that we should try to relate the two phenomena, an idea that has been argued most convincingly, in recent times, by the eminent theoretical physicist, Eugene Wigner. (It should be noted, however, that Wigner's latest work on this topic shows a move to the more conventional position that other complex, yet not conscious, systems can also cause wavefunction reduction. We refer, for example, to his article in *Quantum Optics, Experimental Gravity, and Measurement Theory* ed P Meystre and M O Scully [New York: Plenum 1983].)

We must now explain carefully what is involved here. We consider an isolated system, which may be as complicated as we desire, but which must not contain any conscious mind. According to our assumption, such a system is described by a wavefunction which changes with time according to the rules of quantum theory. A conscious observer now makes a measurement of some property of this system, e.g. of the position of a particle. When the result of this measurement enters the mind of the observer, then the wavefunction reduces to the form corresponding to the particular value of the measured quantity. Notice that it is not enough for the conscious observer simply to be *aware* of only the part of the wavefunction corresponding to the observed value. If this was all that happened then we could not be sure that a different observer would see the same value of the observed quantity. We require that the act of conscious observation actually changes the wavefunction. Thus, in our potential barrier experiment (see figures 15 and 16), if an observer sees the right-hand detector as being ON then it must *be* ON, and not in the state of part ON and part OFF. Then another observer will also see that it is ON, and both will agree that the particle has been reflected. In figure 18 we illustrate this difference between the observed and unobserved systems.

The last paragraph shows the first of the reasons noted by Wigner, in the quotation at the start of this section, for believing that mind affects physical things. The second reason he gives is that in all other parts of physics, action and reaction occur together, i.e. if A affects B then B affects A. Thus, since the physical world clearly affects the conscious mind, we expect the converse to apply.

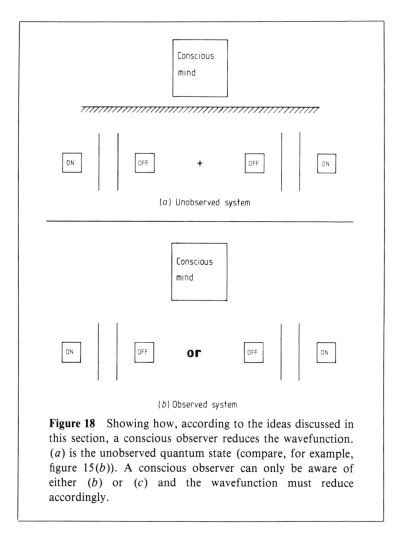

Figure 18 Showing how, according to the ideas discussed in this section, a conscious observer reduces the wavefunction. (*a*) is the unobserved quantum state (compare, for example, figure 15(*b*)). A conscious observer can only be aware of either (*b*) or (*c*) and the wavefunction must reduce accordingly.

To be quite fair, the assertion we have made here, that conscious minds change the wavefunction, is not absolutely necessary. It would, presumably, be possible that the act of observation does not change the system, but that conscious observers somehow communicate with each other so that they all 'see' the same thing. Such an interpretation of quantum theory is possible and we shall discuss it further in §4.5.

We return to our assumption that only conscious observers can reduce wavefunctions, and must now comment on how utterly outrageous such a statement really is. To see this we might suppose that the detectors in the potential barrier experiment are photographic plates. What we are saying is that they are in a state of 'perhaps blackened but perhaps not' until they have been observed by a conscious mind, which may, of course, be years after the event. Indeed, nothing ever really happens, e.g. no particle ever decays, except through the intervention of a conscious mind. We are not quite in the situation of denying external reality—which possibility we considered and rejected in §1.2—but we are denying that the external world possesses the properties we observe, until we actually observe them. This is a picture of reality that we find hard to accept.

The paradox of 'Schrödinger's cat' is an example of the sort of problem we can get into here. We suppose, for example, that the right-hand detector in our potential barrier experiment is a trigger that fires a gun and kills a cat as soon as a particle reaches it. After one particle has passed through the apparatus the wavefunction thus contains a piece in which the cat is dead and a piece in which the cat is alive. Only if the cat is conscious can we say that one of these represents the truth. What however could we say if the cat were asleep? If, on the other hand, a cat is not conscious, or if we used instead a being or a thing that is not conscious, then it remains in a state of being part-dead/part-alive until some conscious observer forces the wavefunction to go to one state or the other. Like Schrödinger himself we probably consider this an unlikely picture of reality.

The assumption we are considering appears even more weird when we realise that throughout much of the universe, and indeed throughout all of it at early times, there were presumably no conscious observers. Thus the wavefunction did not reduce, and all the possibilities inherent in the development of the wavefunction since the beginning of time would have persisted until the first conscious observers appeared. Even worse are the problems we meet if we accept the modern ideas on the early universe in which quantum decays (of the 'vacuum', but this need not trouble us here) were necessary in order to obtain the conditions in which conscious observers could exist. Who, or what, did the observations necessary to create the observers?

The only possibility here seems to be that observation, indeed conscious observation, can be made by 'minds' outside the physical universe. Such is one of the traditional roles of God and/or gods. This is the realm of theology; a realm into which we shall, with trepidation, enter briefly in the next section.

Before we close this section, however, there is one obvious question we must ask. Since we have suggested that consciousness might offer a possible, even if unlikely, solution to a problem of physics, can physics help with the problem of the nature of consciousness? Again the answer may well be that it cannot, but the issue is certainly being discussed. The fact that quantum theory frees physics from the rigid causality of classical mechanics is an obviously immediately relevant fact. There just seems to be more room for ideas like free will in a quantum world than in a classical one. Already quantum tunnelling, as described in §1.3, has been used to explain certain processes in the nervous system—see, for example, Walker, *International Journal of Quantum Chemistry* **11** 103 (1977). (We should, however, be cautious here. There is a big difference between the idea of freedom to choose, where the choice is presumably made by rational thought, and the apparent randomness of quantum theory, so a relation between the two, though possible, is not obvious.) It is also natural to try to associate the very non-local nature of wavefunctions with the similar lack of locality of 'thoughts', etc. For some discussion along these lines, and for other references, we refer to the article by Stapp, 'Consciousness and Values in the Quantum Universe', *Foundations of Physics* **15** 35 (1985).

4.4   God as the conscious observer

Quantum theory offers at least two possible roles for a 'God', where we use this term for a being that is non-physical, non-human, in some sense superhuman, and is conscious.

The first role is to make the 'choices' that are required whenever a measurement is made that selects from a quantum system one of the possible outcomes. Such a God would remove the indeterminacy from the world by taking upon himself those decisions that are not forced by the rules of physics. Although expressed in non-traditional terms, this is reasonably in accordance with the accepted role of a God. He would be very active in all aspects of the world, and would be totally omnipotent within the prescribed limits. Prediction of his behaviour from the laws of physics would be impossible (note that we are not permitting any hidden variables in this chapter), although from both the theological and the scientific viewpoint we would want to believe that there were *reasons* for at least some of the choices; otherwise we would be back with random behaviour and the God would not have played any part. It is interesting to note that this role might even permit 'miracles', if we were to regard these as events so highly unlikely that they would be effectively impossible without very specific, and unusual, 'divine' choice. For example, according to quantum theory, there must be a small, but non-zero, probability that if I run into a wall, then I will pass right through it. This is a special case of the potential barrier experiment and the wavefunction on the left-hand side, corresponding to transmission, is never quite zero. Then, however small the probability for transmission might be, a God would be able to select it as the outcome, if he so chose.

The second possible role for a God to play in quantum theory is more relevant to our principal topic. God might be the conscious observer who is responsible for the reduction of wavefunctions. Whether, in addtion, he also decides the outcome of his observations, as in the above paragraph, or whether this is left to chance is not important here. What is important is the fact that God must be selective—he must not reduce all wavefunctions automatically, otherwise we meet the same problem that we met when discussing modifications to the Schrödinger equation in §3.7: the reduction that is required depends on the observation that we are going to make. If, for example, a reduction to figure 16 is made, then there will be no possibility of interference, whereas a human observer might decide to do the interference experiment.

It is therefore necessary that the God who reduces wavefunctions, and so allows things to happen in the early universe, in particular things that might be required in order for other conscious observers to exist, should know about these other observers and should know what they intend to measure. God must in some way be linked to human consciousness. There is nothing obviously revolutionary in all this (except perhaps the fact that it is discussed in a book on physics), although it perhaps should be mentioned that John Polkinghorne, one-time Professor of Theoretical Physics of

Cambridge and now an Anglican priest, is clearly unhappy with this role for God (see p 67 of the book *The Quantum World* mentioned in the bibliography).

Whether expressed in theological terms or not, the suggestion that conscious minds are in some way connected and that they might even be connected to a form of universal, collective consciousness appears to be a possible solution to the problem of quantum theory. It is not easy to see what it might mean, as we understand so little about consciousness. That there are 'connections' of some sort between conscious mind and physical matter is surely implied by the fact that conscious decisions have effects on matter. Thus there are links between conscious minds that go through the medium of physical systems. Whether there are others, that exploit the non-physical and presumably non-localised nature of consciousness, it is not possible to say. Some people might wish to mention here the 'evidence' for telepathy and similar extra-sensory effects. Such evidence, however, is perhaps better left out of the discussion until it becomes more convincing.

Another way of expressing some of these ideas is to say that there are 'holistic' aspects to quantum physics. Most physics (in particular, all of classical physics) is 'reductionist', in that it tries to explain things in terms of smaller, simpler, localised, entities. It always wants to take things apart and consider their constituents. Quantum theory suggests that this approach will not always be satisfactory. Wavefunctions are very non-local things (we shall see more of the implications of this, and also show how it has been verified experimentally, in Chapter Five).

All this is very vague, and it is certainly not original. Nevertheless, it is remarkable that such ideas should arise from a study of the behaviour of the most elementary of systems. That such systems point to a world beyond themselves is a fact that will be loved by all who believe that there are truths of which we know little, that there are mysteries seen only by the mystics, and that there are phenomena inexplicable within our normal view of what is possible. There is no harm in this—physics indeed points to the unknown. The emphasis, however, must be on the unknown, on the mystery, on the truths dimly glimpsed, on things inexpressible except in the language of poetry, or religion, or metaphor. There can be no support here for the simplistic view of reality that finds its expression in fortune telling or astrological tables, or even in the theological dogmas that are so often the cause of controversy. This cautionary note is well expressd by Bernard d'Espagnat: 'How could we believe that everyday language is able to express otherwise than symbolically the truth about Being itself if it is not able to express the truth about such a trivial object as an atom, except metaphorically?' (*In Search of Reality*, p 109)

4.5 The many-worlds interpretation

In 1957 H Everett III wrote an article entitled '"Relative State" Formulation of Quantum Mechanics' (*Reviews of Modern Physics* **29** 454) which introduced what has become known as the 'many-worlds' interpretation of quantum theory. He began by noting that the orthodox theory requires wavefunctions to change in two distinct ways; first, through the deterministic Schrödinger equation and, secondly, through measurement, which causes the reduction of the wavefunction to a new wavefunction which is not uniquely determined. It is this second type of change that causes problems; what is a 'measurement'?, what are the non-quantum forces that cause it?, how can it occur instantaneously over large distances?, etc. Everett was in fact motivated in his work by yet another problem: he was interested in applying quantum theory to the whole universe, but how could he then have an 'external' observer to measure anything?

The solution that Everett proposed to the problems of wavefunction reduction was to say simply that it does not happen. Any isolated system can be described by a wavefunction that changes only as prescribed by the Schrödinger equation. If this system is observed by an external observer then, in order to discuss what happens, it is necessary to incorporate the observer into the system, which then becomes a new isolated system. The new wavefunction, which now describes the previous system plus the observer, is again determined for all times by the Schrödinger equation.

To help us understand what this means we shall put it into symbolic form. To this end we return to the barrier experiment, in particular as this was discussed in §3.5. We write the wavefunction, after interaction with the barrier, in the form:

$$P_R(W^\rightarrow D_R^{ON} D_L^{OFF}) + P_T(W^\leftarrow D_R^{OFF} D_L^{ON}).$$

This is not really as complicated as it might appear. The Ws describe the particle, with the arrows indicating the direction, and the Ds the two detectors. The first bracket is then the wavefunction of the reflected wave and the second that of the transmitted wave. Each of these wavefunctions is taken to be 'normalised' so that it corresponds to one particle. Then the P_R and P_T are the parameters that give the magnitudes of the two wavefunctions. The squares of these numbers give the probability for reflection and transmission respectively. We notice that this wavefunction correctly describes the correlations between the states of the detectors and those of the particle, e.g. that if the right-hand detector is ON then the particle has been reflected, etc. This correlation exists because, as noted in §3.4, the wavefunction is not simply a product (in fact in this case it is the sum of two products).

According to the orthodox interpretation of quantum theory such a wavefunction reduces, on being observed, to

$$W^{\rightarrow} D_R^{ON} D_L^{OFF} \qquad \text{with probability } P_R^2$$

or to

$$W^{\leftarrow} D_R^{OFF} D_L^{ON} \qquad \text{with probability } P_T^2.$$

(See figure 15.)

In the interpretation due to Everett, however, this reduction does not occur. The true reality is always expressed by the full wavefunction containing both terms. This is all very well, we are saying, but did we not convince ourselves previously that the reduction had to occur; that deterministic theories are not adequate to describe observation? We certainly did, so we must examine the argument. It relied on the fact that *we*, or more properly *I*, do not see both pieces of the wavefunction. To me, either reflection or transmission has occurred, not both. Clearly then, in order to understand what is happening, it is necessary to introduce ME into the experiment and to include ME in the wavefunction. Although my wavefunction is very complicated the only relevant part for our purpose here is whether I am aware of reflection or transmission. We denote these two states of myself by ME^{refl} and ME^{trans} respectively. Thus the complete wavefunction, according to Everett, is:

$$P_R(WDD)^{\rightarrow} ME^{refl} + P_T(WDD)^{\leftarrow} ME^{trans}$$

where we have simplified the notation in an obvious way. Notice that again the wavefunction contains the correct correlations: if the particle is transmitted then I have observed transmission, etc.

Previously we argued (e.g. in §4.1) that, since we are only aware of one possibility, one of the terms in the above expression must be eliminated. Everett would argue instead that there are two MEs, both conscious but unaware of each other. Thus, through my observation of what happens in the barrier experiment, I have split the world into two worlds, each containing one possible outcome of the observation.

Similar considerations apply to other types of observation. In all cases the Everett interpretation requires that *all possible outcomes* exist. Whenever a measurement is made we can think of the world as separating into a collection of worlds, one for each possible result of the measurement. It is through this way of thinking that the name 'many worlds' has arisen. Such a name was not, however, in the original Everett paper, and in some ways it is misleading. The key point of this way of interpreting quantum theory is that measurements are not different from other interactions; nothing special, like wavefunction reduction, happens when a measurement is made; everything is still described, in a deterministic way, by the Schrödinger equation.

How can we reconcile this with our previous belief that measurements were special? The previous argument was basically as follows:

> I am only aware of one outcome of a measurement, therefore there is only one outcome.

Now we would argue differently:

> I am only aware of one outcome of a measurement because the ME that makes this statement, is the ME associated with one particular outcome. There are other MEs, which are associated with different terms in the wavefunction, and which are aware of different outcomes. The wavefunction given above for the barrier experiment illustrates this: both of the terms exist, there are two MEs but they are not aware of each other.

It will be seen that, from the point of view of the many-worlds interpretation, the 'error' we made earlier was that we inserted a tacit assumption that our minds were able to look at the world from outside, and hence to conclude from our certainty of a particular result that the other results had not occurred.

The 'branching' of the world into many worlds is therefore an illusion of the conscious mind. The reality is a wavefunction which always contains all possible results. A conscious mind is capable of demanding a particular result (this is what we mean by making an observation) and thereby it must select one branch in which it exists. Since, however, all branches are equivalent, the conscious mind must split into several conscious minds, one for each possible branch.

Is this then the answer to the problem of reality in the quantum world? At first sight it appears more satisfactory than our previous ideas where consciousness seemed to have to affect wavefunctions; now this is not required. Nevertheless the general view of the theoretical physics community has been to reject the many-worlds interpretation. This of course is not in itself a strong argument against it, particularly when we realise that many writers have rejected it on grounds that suggest they have failed to understand it. Here I should admit that the above discussion was an attempt to describe what I think is the most plausible form of the Everett interpretation. The original paper, and others mentioned in the bibliography, contain mainly the formalism of orthodox quantum theory with little comment on the interpretation.

It is probably fair to say that much of the 'unease' that most of us feel with the Everett interpretation comes from our belief, which we hold without any evidence, that our future will be unique. What I will be like at a later time may not be predetermined or calculable (even if all the initial information were available), but at least I will still be one 'I'. The many-worlds interpretation denies this. For an example to illustrate this lack of uniqueness (some would say rather to show how silly it is) we might return to the barrier experiment and suppose that the right-hand detector is attached to a gun which shoots, and kills, me if it records a particle. Then after one particle has passed through the experiment, the wavefunction would contain a piece with me alive and a piece with me dead. One 'I' would certainly be alive, so we appear to have a sort of Russian roulette, in which we cannot really lose! Indeed, since all 'aging' or 'decaying' processes are presumably quantum mechanical in nature, there is always a small part of the wavefunction in which they will not have occurred. Thus, to be completely fanciful, immortality is guaranteed—*I* will always be alive in the only part of the wavefunction of which *I* am aware!

It is important to realise that the fact that another observer does not see two 'I's is not an argument against this interpretation. As soon as YOU, say, interact with me so that you can discover whether I am alive or dead, you become two YOUs, for one of which I am dead and the other I am alive. In wavefunction language, using the previous notation, we would have:

$$P_R(WDD)^- \text{ME}^{\text{refl}} \text{YOU}^1 + P_T(WDD)^- \text{ME}^{\text{trans}} \text{YOU}^2.$$

Neither of the two YOUs is aware that there are two MEs.

Two final remarks in favour of the many-worlds interpretation should be made here. It has long been known that, for many reasons, the existence of 'life' in the universe seems to be an incredible accident, i.e. if many of the parameters of physics had been only a tiny bit different from their present values then life would not have been possible. Even within the framework of 'design' it is hard to see how everything could have been correct. However, it is possible that most of the parameters of physics were fixed at some early stage of the universe by quantum processes, so that in principle many values were possible. In a many-worlds approach, anything that is possible happens, so we only need to be sure that, for some part of the wavefunction, the parameters are correct for life to form. It is irrelevant how improbable this is, since, clearly, we live in the part of the wavefunction where life is possible. We do not see the other parts. Thinking along these lines is referred to as using the *anthropic principle*; for further discussion we refer to articles listed in the bibliography.

The other remark concerns the origin of the observed difference between past and future, i.e. the question of why the world exhibits an asymmetry under a change in the direction of time when all the known fundamental laws of physics are invariant under such a change. One aspect of this asymmetry is psychological: we remember the past but not the future. (Note that it is because of this clear psychological distinction between past and future that we sometimes find it hard to realise that there is a problem here, e.g. it is possible to fool ourselves that we have derived asymmetric laws, like that concerning the increase of entropy, from laws that are symmetric.) The many-worlds interpretation gives an obvious explanation of this psychological effect: my conscious mind has a unique past, but many different futures. Each time I make an observation my consciousness will split into âs many branches as there are possible results of the observation. Some readers may wish to note that this might allow vague, shadowy, probabilistic, 'glimpses' into the future—thus, a prophecy is likely to be fulfilled, but only for one of the future MEs.

4.6   Summary of Chapter Four

We have considered the possibility that measurement or observation, in the sense required by quantum theory, can only be made by conscious observers. All systems not containing such observers are assumed to be described by a wavefunction which satisfies the rules of quantum mechanics and, in particular, is not reduced. The act of observation by a conscious mind is, however, not describable by quantum theory and does cause wavefunction reduction.

Considerations of those parts and epochs of the universe not observed by conscious beings similar to ourselves led us to speculate on the need for a consciousness not so intimately associated with physical objects, a being to which the name of God appears to be not inappropriate.

We have also introduced a completely different interpretation of quantum theory in which wavefunction reduction does not happen. In this Everett, or many-worlds, interpretation, all possible results of observations exist simultaneously, but conscious minds are only aware of one result. No convincing reasons for rejecting this view appear to exist, though it is very contrary to intuition.

This chapter has been very vague since the topics we have discussed can be neither defined or measured. In the next chapter we shall return to 'real' physics, and even to actual experimental results. We shall not, however, meet anything to undermine our conviction that the content of the present chapter has a relevance to physics that needs to be explored further and that may well prove to be monumental in its significance. In concluding his article, referred to above, Stapp writes: 'For science has already given man the power to solve his major physical problems. The critical remaining problems lie in the sphere of the intellect. Here the dominant influence is the force of ideas. But a shift in the scientific conception of man ... must inevitably deflate egocentric values and enhance the sense of harmonious enterprise with others, and with nature, in the creative unfoldment of new wonders.'

Is this too fanciful?

Chapter Five

Hidden Variables and Non-locality

5.1 Review of hidden-variable theories

In §1.3 we saw that it is possible to repeat an experiment several times, under apparently exactly the same conditions, and yet obtain different results. In particular, for example, we could direct identical particles, all with the same velocities, at identical potential barriers, and some would be reflected and some transmitted. The initial conditions would not uniquely determine the outcome. Quantum theory, as explained in Chapter Two, accepts this lack of determinism; knowledge of the initial wavefunction only permits probabilistic statements regarding the outcome of future measurements.

Hidden-variable theories have as their primary motivation the removal of this randomness. To this end they regard the 'apparently' identical initial states as being, in reality, different; distinguished by having different values of certain new variables, not normally specified (and therefore referred to as 'hidden'). The states defined in quantum theory would not correspond to precise values of these variables, but rather to certain specific averages over them. In principle, however, other states, which do have precise values for these variables, could be defined and with such initial states the outcome of any experiment would be uniquely determined. Thus determinism, as understood in classical physics, would apply to *all* physics. Particles would then have, at all times, precise positions and momenta, etc. The wavefunction would *not* be the complete description of the system and there would be the possibility of solving the problems with wavefunction reduction which we met in Chapter Three. This latter fact is, to me at least, a more powerful motivation than the desire for restoration of determinism.

Any satisfactory hidden-variable theory must, of course, agree with experimental observations and therefore, in particular, with all the verified predictions of quantum theory. Whether it should agree *exactly* with quantum theory, or whether it might deviate from it to a small degree, while still remaining consistent with experiment, is an open question. The normal practice seems to have been to seek hidden-variable theories for which the agreement is exact. A hidden-variable theory will, of course, tell us *more* than quantum theory tells us—for example, it tells us which particles will pass through a given barrier. What we require is that it gives the same, or very nearly the same, results for those quantities that quantum theory can predict.

There have been, and still are, many physicists who would regard the question of the possiblity of such a hidden-variable theory, agreeing in all measurable respects with quantum theory, as being an unimportant issue. Readers who are still with us, however, are presumably convinced that the quest for reality is meaningful, so they will take a different view. The question *is* interesting and worthy of our attention. Indeed, there are even pragmatic grounds for pursuing it: different explanations of a set of phenomena, even though they agree for all presently *conceivable* experiments, may ultimately themselves suggest experiments by which they could be distinguished. There is also the hope that better understanding of quantum theory might help in suggesting solutions to some of the other unsolved problems of fundamental physics.

The subject of hidden-variable theories was for many years dominated by an alleged 'proof', given by von Neumann in 1932 (in his book *Mathematische Grundlagen der Quantenmechanik* [Berlin : Springer], English translation published by Princeton University Press, 1955), that such theories were impossible, i.e. that no hidden-variable, deterministic, theory could agree with all the predictions of quantum theory. The proof was simple and elegant; its mathematics, though subject to much scrutiny, could not be challenged. However, the mathematical theorem did not really have any relevance to the physical point at issue. The reason for this lay in one of the assumptions used to prove the theorem. We shall give a brief account of this assumption in the following paragraph. Since this account is rather technical and not used in the subequent discussion, some readers may prefer to omit it.

Let us suppose that two quantities, call them X and Y, can be separately measured on a particular system, and that it is also possible to measure the sum of the two quantities, $X + Y$, directly. Then the assumption was that the average value of $X + Y$, over any collection of identical systems, i.e. any ensemble, was equal to the average value of X plus the average value of Y. Since, in general, the variable $X + Y$ is of a different kind, measured by a different apparatus, from either X or Y, there is no reason why such an equality should hold. Von Neumann was led to assume it because it happens to be true in quantum theory, i.e. for those ensembles specified by a given wavefunction. In a hidden-variable theory, however, other states, defined by particular values of the hidden variables, can, at least in principle, exist, and for such states the assumption does not have to be true. Although several people seemed vaguely to have realised this problem with von Neumann's theorem, it was not until 1964 that John Bell finally clarified the issue, and removed this theoretical obstacle to hidden-variable theories. The article was published in *Reviews of Modern Physics* **38** 447 (1966).

At this stage we should emphasise that, although hidden variable theories are *possible*, they are, in comparison to quantum theory, extremely complicated and messy. We *know* the answers from quantum theory and then we construct a hidden-variable, deterministic, theory specifically to give these answers. The resulting theory appears contrived and unnatural. It must, for example, tell us whether a given particle will pass through a potential barrier for all velocities and all shapes and sizes of the barrier. It must also tell us the results for any type of experiment; not only for the reflection/transmission barrier experiment of §1.3, but also for the experiment with the mirrors. In the latter case, there can now be no question of interference being the real explanation of what is happening, because a given particle is certainly either reflected or transmitted by the barrier and hence can only follow one path to the detectors. Nevertheless, although it reaches only one of the mirrors, which reflects it to the detectors, the path it follows must be influenced by the other mirror. This is brought about by the introduction of a new 'quantum force' which can act over arbitrarily large distances. This quantum force is constructed in order to give the required results.

For details of all the various hidden-variable theories that are available we refer to the excellent book by Belinfante, *A survey of hidden-variable theories* [Oxford: Pergamon 1973]. Here, we shall only discuss a particular class of such theories; they appear to be the most plausible and are the topic of our next section.

5.2 The pilot wave

I think that conventional formulations of quantum theory, and of quantum field theory in particular, are unprofessionally vague and ambiguous. Professional theoretical physicists ought to be able to do better. Bohm has shown us a way.†

In the very early days of quantum theory, de Broglie, who had been the first to associate a wavefunction with a particle, suggested that, instead of being the complete description of the system, as in conventional quantum theory, the true role of this wavefunction might be to guide the motion of the particles. In such a theory the wavefunction is therefore called a *pilot wave*. The particles would always have precise trajectories, which would be determined in a unique way from the equations of the theory. It is such trajectories that constitute the 'hidden variables' of the theory.

These ideas were not well received; probably they were regarded as a step backwards from the liberating ideas of quantum theory to the old restrictions of classical physics. Nevertheless, and in spite of von Neumann's theorem discussed above, interest in hidden-variable theories did not completely die, and in 1952 David Bohm produced a theory based on the pilot wave idea, which was deterministic and yet gave the same results as quantum theory. It also provided a clear counter-example to the von Neumann theorem.

In Bohm's theory a system at any time is described by a wavefunction *and* by the positions and velocities of all the particles. (Since it is positions that we actually observe in experiments, it is perhaps paradoxical that these are called the 'hidden' variables, in contrast to the wavefunction.) To find the subsequent state of the system, it is necessary first to solve the Schrödinger equation and thereby obtain the wavefunction at later times. From this

† J Bell, *Beables for quantum field theory* (*CERN preprint* TH4035/84, 1984)

wavefunction a 'quantum force' can be calculated. This force is added to the other, classical, forces in the system, e.g. those due to electric charges, etc, and the particle paths are then calculated in the usual classical way by Newton's laws of motion. The quantum force is chosen so that there is complete agreement with the usual predictions of quantum mechanics. What we mean by this is that, if we consider an ensemble of systems, with the same initial wavefunction but different initial positions, chosen at random but with a distribution consistent with that given by the wavefunction, then at any later time the distribution of positions will again agree with that predicted by the new wavefunction appropriate to the time considered. It is beyond the scope of this book to discuss further the technical details, and problems, associated with these considerations.

In comparing the Bohm–de Broglie theory with ordinary quantum theory we note first that, since they give the same results for all quantities that we know how to measure, they are equally satisfactory with regard to experiments. As far as we know both are, in this sense, correct. The former has the added feature of being deterministic, but with our present techniques this is not significant experimentally. The degree to which it is regarded as a conceptual advantage is a matter of taste.

A much more important advantage of the hidden-variable theory is that it is precise. It is a theory of everything; no non-quantum 'observers' are required to collapse wavefunctions since no such collapse is postulated. All the problems of Chapter Three disappear.

In connection with this last observation, we should note two points. First, it may be asked how we have been able to remove the requirement for wavefunction collapse when, in Chapter Three, we appeared to find it necessary. The answer lies in the fact that, whereas previously the wavefunction was the complete description of the system, so that there was no place for the difference between transmission or reflection (for example) to show other than in the wavefunction, now that we have additional variables to describe the system this is no longer the case. The wavefunction can be identical for both transmission and reflection, since the difference now lies in the hidden variables, in particular in the positions of the particle.

Secondly we should note a reservation to the remark above that the two theories always agree. Readers may indeed be wondering how this can be, when in one case we have wavefunction collapse but not in the other. The answer lies in the fact that wavefunction collapse only happens in the orthodox interpretation when macroscopic measuring devices are involved. It is only when the wavefunction can be written as the sum of macroscopically different pieces that some of them are dropped in the process of reduction. Now the difference between keeping all the pieces, as in the Bohm–de Broglie theory, and dropping some of them, as in the orthodox theory, is only significant experimentally if they can be made to interfere. However, such interference can only occur if the pieces can be made identical, which as we have seen (§3.6 and Appendix 6) is so unlikely for macroscopic objects as to be effectively impossible. The two theories are experimentally indistinguishable because macroscopic processes are not reversible. Nevertheless we should emphasis that, where interference can in principle occur, it is indeed observed. There is no positive evidence that wavefunction reduction actually happens, so, especially in view of the problems of Chapter Three, theories that do not require it have a real advantage.

Given this fact it is perhaps rather remarkable that hidden-variable theories are not held in high regard by the general community of quantum physicists. Why is this so? More importantly, are there any good reasons why we should be reluctant to accept them?

We have already hinted at some of the possible answers to the first question. The many successes of quantum theory created an atmosphere in which it became increasingly unfashionable to question it; the argument between (principally) Bohr and Einstein on whether an experiment to violate the uncertainty principle could be designed was convincingly won by Bohr (as the debate moved into other areas the outcome, as we shall see, was less clear); the elegance, simplicity and economy of quantum theory contrasted sharply with the contrived nature of a hidden-variable theory which gave no new predictions in return for its increased complexity; the whole hidden-variable enterprise was easily dismissed as arising from a desire, in the minds of those too conservative to accept change, to return to the determinism of classical physics; the significance of not requiring wavefunction reduction could only be appreciated when the problems associated with it had been accepted and, for most physicists, they were not, being lost in the mumbo-jumbo of the 'Copenhagen' interpretation; this interpretation, due mainly to Bohr, acquired the status of a dogma. It appeared to say that certain questions were not allowed so, dutifully, few people asked them.

With regard to the second of the questions raised above (namely, are there any good reasons for rejecting the hidden-variable approach?), it has to be said that the picture of reality presented by the Bohm—de Broglie theory is very strange. The quantum force has to mimic the effects of interference so, although a particle follows a definite trajectory, it is affected by what is happening elsewhere. The reflected particle in figure 2 somehow 'knows about' the left-hand mirror, though its path does not touch it; similarly, the particle that goes through the upper slit in the double slit experiment shown in figure 13 'knows' whether the lower slit is open or not. This 'knowledge' arises through the quantum force which can apparently operate over arbitrarily large distances. To show in detail the effect of this force we reproduce in figure 19 the particle trajectories for the double-slit experiment as calculated by Philippidis *et al* (*Il Nuovo Cimento* **52B** 15, 1979). We remind ourselves that, if we are to accept the Bohm theory, then we must believe the particles really do follow these peculiar paths. Particles have become real again, exactly as in classical physics, the uncertainty has gone, but the price we have paid is that the particles behave very strangely!

Another, perhaps mainly aesthetic, objection to hidden-variable theories of this type is that, without wavefunction reduction, we have something similar to the many-worlds situation, i.e. the wavefunction contains all possibilities. Unlike the many-worlds case, these are not realised, since the particles all follow definite, unique, trajectories, but they are nevertheless present in the wavefunction—waiting, perhaps, one day to interfere with what we think is the truth! Thus, in our example discussed in Appendix 2, both scenarios act out their complete time development in the wavefunction. It is all there. The real, existing wavefunction of the universe is an incredibly complicated object. Most of it, however, is irrelevant to the world of particles, which are the things that we actually observe.

The unease we feel about such apparent redundancy can be made more explicit by expressing the problem in the following way: the pilot wave affects the particle trajectories, but the trajectories have no effect on the pilot wave. Thus, in the potential barrier experiment, the reflected and transmitted waves exist and propagate in the normal way, totally independent of whether the actual particle is reflected or transmitted. This is a consequence of the fact that the wavefunction is calculated from the Schrödinger equation which does not mention the hidden variables. It is a situation totally contrary to that normally encountered in physics, where, since the time of Newton, we have become accustomed to action and reaction occurring together.

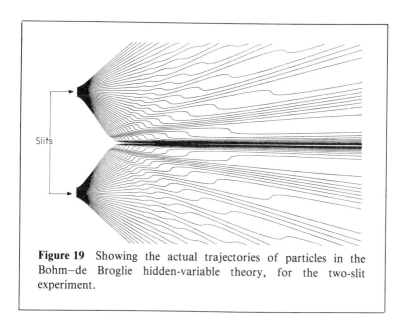

Figure 19 Showing the actual trajectories of particles in the Bohm–de Broglie hidden-variable theory, for the two-slit experiment.

To return to more concrete matters, there is one problem with the Bohm theory which we must mention, but into which we cannot go far, namely, that it is not compatible with special relativity. This is also true of ordinary quantum theory, but the problem with hidden-variable theories may be more serious. Attempts to combine quantum theory with relativity go through the highly successful quantum field theory. Recently, Baumann (Graz preprint, *Interpretation of macroscopic quantum phenomena*, 1984) and Bell (CERN preprint referred to at the start of this section) have extended the Bohm–de Broglie ideas to quantum field theory. It is, however, unclear whether the resulting theory is, or can be made, consistent with special relativity.

Finally, we refer again to the quantum force. This is unlike all the other forces in physics because it does not seem to be caused by the exchange of particles. Here we remind readers that the four known types of force are:

electromagnetic, caused by the exchange of photons;
strong, caused by the exchange of gluons (slightly indirectly);
weak, caused by the exchange of W and Z bosons;
gravitational, caused by the exchange of gravitons. (These have not actually been observed, but this is not surprising in view of the weakness of the gravitational coupling. Few theoretical physicists would doubt their existence.)

The quantum force also has the remarkable property of acting regardless of distance. At the end of his paper on von Neumann's theorem (referred to above), John Bell noted how unnatural and unphysical this was, and raised the question of whether it was an essential feature of all hidden-variable theories, or whether it was merely a defect of those that were available at the time. Before publication of the paper he had already given an affirmative answer: no local hidden-variable theory can agree with all the predictions of quantum theory. We defer discussion of this to §5.4, after we have introduced, in the next section, a class of experiments which forcibly demonstrate the non-locality which is intrinsic to quantum theory.

5.3   The Einstein–Podolsky–Rosen thought experiment

In 1935, Einstein, Podolsky and Rosen published a paper entitled 'Can Quantum-Mechanical Description of Physical Reality Be Considered Complete?' (*Physical Review* **47** 777), which has had, and continues to have, an enormous influence on the interpretation problem of quantum theory. In this paper, they proposed a simple thought experiment and analysed the implications of the quantum theory predictions for the outcome of the experiment. These made explicit the essentially non-local nature of quantum theory and, according to the authors, proved that the theory must be incomplete, i.e. that a more complete (hidden-variable) theory exists and might one day be discovered. Much later, as we discuss in the next section, John Bell carried the analysis considerably further and showed that no *local* hidden-variable theory could reproduce all the predictions of quantum theory. Naturally this work prompted experimentalists to turn the thought experiments into real experiments, and so check whether these predictions are correct, or whether the actual results deviated from them in such a way as to permit the existence of a satisfactory local theory. These experiments, which we discuss in §5.5, beautifully confirm quantum theory.

We shall refer to the general class of experiments with the same essential features as that proposed by Einstein, Podolsky and Rosen as EPR experiments. The orginal work is sometimes called the EPR paradox, or the EPR theorem.

The particular EPR experiment that we shall describe is somewhat different from the original, but is more suited to our later discussion. We consider the situation shown in figure 20, in which a particle with zero spin at rest in the laboratory decays spontaneously into two, identical, particles, each with spin 1/2. These particles, which we call A and B respectively, will move apart with velocities that are equal in magnitude and opposite in direction. (This ensures that their momenta add to zero so that the total momentum, which was initially zero, is conserved.)

The experiment now consists of measuring the spin components of the two particles in any particular directions—in fact, for simplicity, we consider only directions perpendicular to the direction of motion. Thus we have an apparatus that will measure the spin component of particle A in a direction we can specify by the angle a. Similarly we have an apparatus to measure the spin component of B in a direction specified by some angle b. The full experiment is illustrated in figure 21. The form of the apparatus used to measure the spin is irrelevant for our purpose, but in order to demonstrate that the measurement is possible we could consider the case in which the spin 1/2 particles are charged, e.g. electrons. In that case the particles would have a magnetic moment which would be in the same direction as the spin. Then to measure the spin along a specific direction we could have a varying magnetic field in that direction which would deflect the electron, up or down according to the value of the spin component.

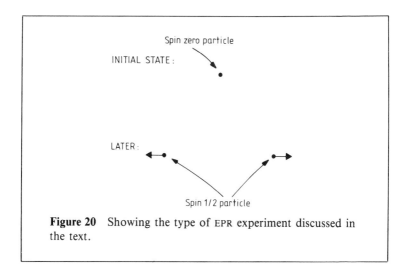

Figure 20   Showing the type of EPR experiment discussed in the text.

In order to discuss the form of the results we must digress a little to think about spin. We first recall, from the earlier discussion of spin in §3.7 (also Appendix 8), that a measurement of a spin component of a spin $1/2$ particle in any given direction will always give a value either $+1/2$ or $-1/2$, i.e. the spin is always either exactly along the chosen direction or exactly contrary to it. Suppose, for example, that we know the particle has a spin component $+1/2$ in a particular direction (see figure 22). Whereas according to classical mechanics we would obtain some value in between $+1/2$ and $-1/2$ for this second measurement, in fact, according to quantum theory, we will obtain either of the two extremes, each with a calculable probability. This probability will depend on the angle between the two directions, and will be such that the average value agrees with that given by classical mechanics. Within quantum theory it will not be possible to predict which value we will obtain in a given measurement; the situation in fact will be very analogous to the choice of reflection or transmission in the barrier experiment of §1.3.

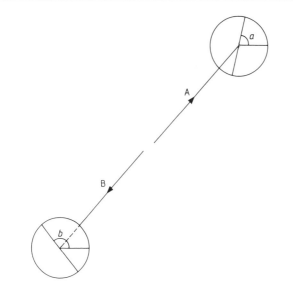

Figure 21 Showing the directions of the spin measurements in the EPR experiment. A and B are the paths of the two electrons and are perpendicular to the plane of the paper. The circles are in the plane of the paper, and the detectors measure the spin projections along lines specified by the angles *a* and *b*, which can be varied by rotating the detectors in the plane of the paper.

Further details of all this are given in Appendix 8. For the following discussion the important fact we shall need to remember is that, in quantum theory, the spin of a particle can have a definite value in only one direction. We are free to choose this direction, but once we have chosen it and determined a value for the spin in that direction, the spin in any other direction will be uncertain. The fact that when we measure the spin in this new direction we automatically obtain a precise value implies that the measurement does something to the particle, i.e. it forces it into one or the other spin values along the new line. This of course is an example of wavefunction reduction about which we have already written much.

The next thing that we need to learn is that the total spin, in any given direction, for an isolated system, remains constant. Readers who know about such things will recognise this as being related to the law of conservation of angular momentum. It is true in quantum mechanics, as well as in classical mechanics; in particular, it is true for individual events and not just for averages, a fact which has been experimentally confirmed.

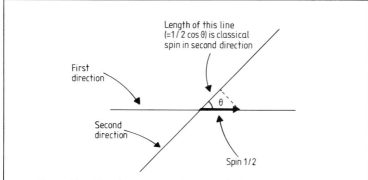

Figure 22   Showing how a spin of $+1/2$ in a particular direction projects to a particular value $(1/2 \cos \theta)$ in a second direction according to classical physics. In quantum theory, the spin along the new direction has to be $+$ or $-1/2$, with probabilities such that the average value is equal to the classical prediction.

What does this law tell us in our EPR experiment? To answer this we recall that the initial spin is zero in any direction (because we chose to do the experiment starting with a particle of zero spin). Thus if we measure the spin of one of the produced particles along a given direction, then we know, even without doing a measurement, the spin of the other in the same direction. To be specific, if we measure the spin of A, in the direction specified by the angle a, and obtain the value $+1/2$, then we know that a subsequent measurement of the spin of B along the same direction, i.e. with $b = a$, will give the value $-1/2$. In other words, B has a definite value of its spin along this particular direction.

Einstein, Podolsky and Rosen then argued that, since a measurement of the spin of A cannot affect particle B, which in principle could be millions of miles away, B must have this definite value of spin even if the spin of A is not measured (of course, we would not know what it was, but this does not affect the argument). We are then able to make exactly the same argument for *any* direction, i.e. for any choice of the angle *a*, and hence conclude that particle B has a definite spin in all directions simultaneously, in direct contradiction to what is allowed in quantum theory. Thus the authors concluded that the quantum theory description must be incomplete.

It is worth recording here more precisely some of the details of the argument given in the original EPR paper. The following two criteria were introduced:

(i) If, without in any way disturbing a system, we can predict with certainty (i.e. with probability equal to unity) the value of a physical quantity, then there exists an element of physical reality corresponding to this physical quantity.

(ii) Every element of the physical reality must have a counterpart in the physical theory.

From the above discussion, the first criterion can then be used to show that the spin of B in any direction is an element of physical reality. It does not have a place in the quantum theory description of spin, which must therefore, according to the second criterion, be incomplete.

The answer that quantum theory itself makes to this charge is that in the EPR experiment it is just not true to say that measurement of the spin of A does nothing to B. In fact it affects B just as surely as if the measurement were made on B itself. We have seen that this follows through the law of spin conservation. It is instructive to see explicitly how it happens in the framework of quantum theory. This will require us to enter into a little symbolism, but as it is really fairly simple, maybe this can be forgiven.

We suppose that the wavefunction for particle A with spin $+1/2$ or $-1/2$ in the direction denoted by the angle *a*, is written as $V^+(a)$ or $V^-(a)$, respectively. The wavefunctions for particle B are similarly written as $W^+(b)$ and $W^-(b)$. We are here ignoring that part of the wavefunction that tells us about the position and velocity of the particles since it does not enter into our discussion.

In order to write the wavefunction for the system of two particles we choose the same value for *a* and *b*. There are then two possible results for the measurements: either we can have $+1/2$ for A and $-1/2$ for B, or vice versa. The complete wavefunction must contain both possibilities, with equal probability. In fact it has the form

$$V^+(a)\,W^-(a) - V^-(a)\,W^+(a).$$

This may look a little confusing, but it can easily be seen to have the properties we require. First we note that, although we have used a particular direction, denoted by a, in both wavefunctions, any direction will do for a, and it turns out that the wavefunction does not change if we change the value of a (this is not 'obvious'; we prove it in Appendix 8, but readers who do not want such details can just accept it). The minus sign between the two terms may come as a surprise. It is there because the wavefunction corresponds to spin zero in *all* directions, not just in the direction given by the angle a.

We can now see how a measurement on A affects the wavefunction of B, and how spin conservation is ensured. If, for example, we find the value $+1/2$ then we select the first of the two terms in the wavefunction. This guarantees that particle B is in the state with spin $-1/2$ in the chosen direction. The crucial point is that, as we have noted before, particles which have once interacted are never really independent. Doing something to one of them can affect the wavefunction of the other, regardless of how large their separation might be.

Is this reasonable? Can we really accept this lack of 'locality'? It is certainly very uncomfortable. For example, we could make our measurements long after A and B have separated, so that B might have travelled to some distant planet. We like to believe that we can consider an isolated system that will not be affected by something that might, unbeknown to us, be happening on another planet. The degree of our discomfort will depend upon the degree of reality which we ascribe to the wavefunction. The so-called 'Copenhagen' interpretation, mainly due to Bohr, essentially denied reality to anything but the results of measurements. This solves the problem by pretending that it does not exist. Since we have committed ourselves to a belief in reality, such an escape is not available to us.

Can hidden variables help us? In hidden-variable theories the wavefunction does not completely specify the system, so we might hope that the change in the wavefunction is only really a change in our knowledge of the system, and that the actual values of the variables that specify, for example, the spin state of particle B do not alter when we measure the spin of A. In the next section we shall rule out such a hope. We shall show that no hidden-variable theory that obeys a reasonable criterion of locality can agree with the predictions of quantum theory, or indeed (§5.5) with experiment.

Note that the Bohm–de Broglie type of hidden-variable theory does not remove the non-locality. Indeed, as Bell remarked, it explains the EPR correlations in the way that Einstein would have least liked. Far from removing the non-locality, it reveals it in a clear way. The quantum force, seen by particle B, is affected by what we choose to measure at the position of particle A.

5.4   Bell's theorem

This theorem, published in 1964 (*Physics* **1** 195), expresses one of the most remarkable results of twentieth century theoretical physics. It exposes, in a clear quantitative manner, the real nature of the conflict between 'common sense' and quantum theory which exists in the EPR type of experiment. As we shall show, the theorem is easy to prove (once one has seen it), but the fact that nobody at the time of the early controversy following the publication of the EPR paper realised that such a result could be found is the real measure of the magnitude of John Bell's achievement.

In order to appreciate properly the meaning of the theorem we must first emphasise an important distinction; one which we have indeed already met. The EPR experiment suggests that measurements on one object (A) alter what we can predict for subsequent measurements on another object (B), regardless of how far apart the objects may be at the time of the measurements. There are two completely different ways of explaining this, namely:

(i) it could be that measurements on A actually have an effect on B, or alternatively,

(ii) it could be that measurements on A only affect our knowledge of the state of B, i.e. they tell us something about B which was in fact already true before the measurement.

The first of these possibilities, which Bell's theorem shows is the case in quantum theory, is totally contrary to the idea of locality. The second, on the other hand, is an everyday occurrence and has no great significance.

As a trivial example illustrating the last remark, we imagine that a box is known to contain two billiard balls, one of which is black and the other white. We then remove one ball, in the dark, and put it on a rocket which flies off into space. At this stage all that we know about the colour of this ball is that there is a 50% chance of its being white, and a 50% chance of its being black (just like a spin in a given direction might have a 50% chance of being either $+1/2$ or $-1/2$). We then look at the ball remaining in the box and if it is black (white) we immediately know that the other ball is white (black). Again this is superficially rather like our experiment with two spin 1/2 particles. However we know that in no sense do we do anything to the distant ball by looking in the box. It already was either white or black. Because of our lack of knowledge, our previous description of it was incomplete. A complete description did however exist, and with such a complete description, the observation of the colour of the remaining ball would clearly have no effect.

The question now is whether such a complete description can exist in the EPR spin experiment, i.e. is it possible that there is a way of specifying the state of particle B such that measurements on A have no effect on B? Bell's theorem allows us to give a negative answer to this question both on the basis of quantum theory, and of experiment (see next section).

It is instructive to see exactly what is involved in the theorem, in particular, how little, so we shall give the proof even though it again involves a small amount of mathematical symbolism. (A simpler form of the theorem, described in terms of the behaviour of people rather than particles, can be found in Appendix 9.)

To begin, we suppose that the spin-measuring apparatus, at each side, is connected to a machine that records the results of the measurements. We arrange that these machines record $+1$ for a spin measurement of $+1/2$ and -1 for a measurement of $-1/2$. Let M and N be the values recorded for the A and B particles respectively. In fact we shall be concerned only with the product of M and N, which we denote by E. The appropriate experimental arrangement is depicted in figure 23. (Note that throughout this section we are making the natural assumption that a measurement gives only one result. Thus we are ignoring the many-worlds possibility. For further discussion of this point see the article of Stapp, 'Bell's theorem and the foundations of quantum physics', *American Journal of Physics* **53** 306 1985.)

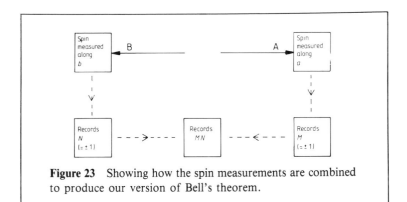

Figure 23   Showing how the spin measurements are combined to produce our version of Bell's theorem.

Not surprisingly, in view of the statistical nature of quantum theoretical predictions, the argument requires us to consider not just one event but many, i.e. the decay of a large number of identical spin zero particles. For each such event we can record a value of E (always + or −1), and we then calculate the average over all events. This will depend upon the orientation of the two spindetectors, which are given by the angles a and b, so we write it as $\langle E(a, b)\rangle$. Thus,

$$\langle E(a, b)\rangle = \text{Average value of } E$$
$$= \text{Average value of } M \cdot N. \qquad (5.1)$$

Clearly this number lies between + 1 and − 1.

Next we introduce the variable H which is supposed to give the required complete description of the two spin 1/2 particles. It is not important for our purpose whether H consists of a single number or several numbers. However, for convenience we shall refer to it as though it were just a single number. When we know the value of H we know everything that can be known about the system. Each event will be associated with a certain value of H and in a number of such events there will be a certain probability for any particular value occurring. If the hidden-variable theory is deterministic (a restriction we shall later drop), then the values of M and N in a given event, and for given angles a and b, are uniquely determined by the value of H.

Now we introduce the assumption of locality which is here expressed by the assertion that the value of M does not depend on b and the value of N does not depend on a. In other words, the value we measure for the spin of the particle A cannot depend on what we choose to measure about particle B, and vice versa. It follows that M depends only on H and a, whilst N depends only on H and b. We express these dependences by writing the values obtained as $M(H, a)$ and $N(H, b)$ respectively. The resulting value of E is then given by

$$E(H, a, b) = M(H, a)N(H, b). \qquad (5.2)$$

For a particular value of H this is a fixed number. Different values of H can occur when we repeat the experiment many times, and the average value of E that is measured will equal the average of $E(H, a, b)$ over these values of H, i.e. the hidden-variable theory predicts

$$\langle E(a, b) \rangle = \text{Average over } H \text{ of } E(H, a, b). \qquad (5.3)$$

At this stage we do not appear to have got very far. Since we do not know anything about the variation of M or N with H, or about the distribution of the values of H, all that we can say about the predicted value of $\langle E(a,b) \rangle$ is that it lies between $+1$ and -1. This of course we already knew.

Now comes the clever part. We consider two different orientations for each of the spin measuring devices. Let these be denoted by the angles a and a' for measurements on A and by b and b' for measurements on B. For a fixed value of H, there are now two values of M and two values of N, i.e. four numbers, each of which is either $+1$ or -1. In table 5.1 we show all possible sets of values for these four numbers. We also show the corresponding values for the quantity $F(H, a, a', b, b')$, defined by

$$F(H, a, a', b, b') = E(H, a, b) + E(H, a', b')$$
$$+ E(H, a', b) - E(H, a, b'). \qquad (5.4)$$

Table 5.1

$M(a)$	+	+	+	+	+	+	+	+	−	−	−	−	−	−	−	−
$M(a')$	+	+	+	+	−	−	−	−	+	+	+	+	−	−	−	−
$N(b)$	+	+	−	−	+	+	−	−	+	+	−	−	+	+	−	−
$N(b')$	+	−	+	−	+	−	+	−	+	−	+	−	+	−	+	−
F	2	2	−2	−2	−2	2	−2	2	2	−2	2	−2	−2	−2	2	2

In all cases this quantity is $+$ or -2, from which it follows that its average value over the (unknown) distribution of H lies between -2 and $+2$. Hence our local hidden-variable theory predicts that the particular combination of results defined by

$$\langle F(a, a', b, b') \rangle = \langle E(a, b) \rangle + \langle E(a', b') \rangle$$
$$+ \langle E(a', b) \rangle - \langle E(a, b') \rangle \quad (5.5)$$

satisfies:

$$-2 < \langle F(a, a', b, b') \rangle < +2. \quad (5.6)$$

This is one form of the Bell inequality.

It is important to realise that locality rather than determinism is the key ingredient of this proof. In order to demonstrate this, we relax the assumption that H determines the values of M and N uniquely, and suppose instead that each value of H determines a probability distribution for M and N. The locality assumption is now a little more subtle. It is that the probability distribution for M does not depend on the value measured for N, and vice versa. To appreciate why this is so we recall that measurement of N cannot tell us anything further about particle A, since H is intended to be the complete description of the state of the system; equally, because of locality, it cannot change the state of A. Hence the probability of obtaining any given value of M does not depend on the value measured for N.

We can then define independent averages of M and N, for each value of H. We denote these by $M^{av}(H, a)$ and $N^{av}(H, b)$. Because of the assumption of independence, the average value of the product of M and N, which we write as E^{av}, is equal to the product of the average values, i.e.

$$E^{av}(H, a, b) = M^{av}(H, a)N^{av}(H, b). \quad (5.7)$$

It is now possible to prepare a table similar to that above with M^{av} replacing M, etc. Instead of taking values of $+$ or -1, these quantities lie somewhere between these limits. It is then quite easy to show that the particular combination defining F, which we now denote by F^{av}, always takes a value that lies between -2 and $+2$. When we then average F^{av} over all values of H we again obtain the Bell inequality.

For a more complete discussion of the circumstances in which the inequality, or various alternative versions of it, can be proved we refer to the review article of Clauser and Shimony listed in the bibliography (§6.5)

The significance of the Bell inequality lies in the fact that, unlike the inequality we found for E, it does not have to be true if the locality assumption is dropped. Indeed it turns out that the inequality is *violated* by the predictions of quantum theory.

Before we discuss these predictions it is interesting to see why quantum theory fails to satisfy the assumptions of the theorem. In quantum theory the full specification of the state is the wavefunction, so this plays the role of the quantity H. We can then define, as above, the averages of M and N over many measurements. However, these averages are not independent; the distribution of values of M depends on what has been measured for N. As a simple illustration of this we note that, with our wavefunction corresponding to total spin zero, the averatge value of M or N measured independently is zero (regardless of the angles a, b). However, in the special case of $a = b$, then we know that M is always opposite to N, so the product is always -1. Thus the average value of MN is -1, which is not the product of the separate averages of M and N.

In general, the quantum theoretical prediction for $\langle E(a, b) \rangle$ depends on the difference between the angles a and b. As we show in Appendix 8 it is given by

$$\langle E(a, b) \rangle = -\cos(a - b). \qquad (5.8)$$

This function is drawn in figure 24. As expected it lies between -1 and $+1$. However, and this is the reason why it leads to a conflict with the Bell inequality, it cannot be factorised into a product of a function of a and a function of b.

The resulting prediction for $\langle F(a, a', b, b') \rangle$ is now easily found. A particularly simple case is when the angles are chosen as in figure 25. Here the violation of the inequality is maximised; each term in F contributing the same amount, $(\sqrt{2})/2$, to the sum. Hence,

$$F = 2\sqrt{2} \simeq 2.83$$

which is in clear violation of the Bell inequality.

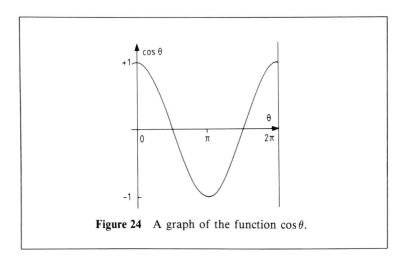

Figure 24 A graph of the function $\cos\theta$.

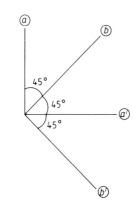

Figure 25 Showing the relative orientations of the spin measurements such that the Bell inequality is maximally violated.

Thus the Bell inequality shows that any theory which is local must contradict some of the predictions of quantum theory. The world can either be in agreement with quantum theory or it can permit the existence of a local theory; both possibilities are not allowed. The choice lies with experiment; the experiments have been done and, as we explain in the next chapter, the answer is clear.

5.5   Experimental verification of the non-local predictions of quantum theory

As we discussed in §2.5, quantum theory has been successfully applied to a truly enormous variety of problems, and its status as a key part of modern theoretical physics, with applications ranging from the behaviour of the early universe and the substructure of quarks to practical matters regarding such things as chemical binding, lasers and microchips, is unquestioned. New tests of such a theory might therefore be seen as adding very little to our knowledge. The reason why, in spite of this, the experiments which we describe here have attracted so much attention is that they test certain simple predictions of the theory which violate conditions (in particular the Bell inequalities) that very general criteria of localisability would lead us to expect.

Following the publication of the first of the Bell inequalities, in 1965, there have been a succession of attempts to test them against real experiments. These experiments are, in fact, quite difficult to do with sufficient accuracy and early attempts, although they generally supported quantum theory, with one exception, were rather inconclusive. We shall therefore confine our discussion to the recent series of experiments which have been performed in France by Aspect, Dalibard, Grangier and Roger.

In all these experiments a particle emits successively two photons in such a way that their total spin is zero. We recall that photons are the particles associated with electromagnetic radiation, e.g. light. They are spin one particles, in contrast to the spin 1/2 particles which we have previously used in our discussion. It is convenient to measure the 'polarisation' of the photons rather than their spin projections. These are related in a way that need not concern us. The only difference we need to note is that in the predicted expression for $\langle E \rangle$ the angle between the various directions has to be doubled, i.e. we find

$$\langle E(a,b) \rangle = -\cos 2(a-b). \qquad (5.9)$$

In the first experiment the spin measurements were carried out in such a way that a particle with spin $+1$ in a chosen direction was deflected into the detector and counted, whereas a particle with spin -1 in the same direction was deflected away from the detector and not counted. The experiment then measured the number of coincident counts, i.e. counts at both sides. Because of imperfections in the detectors it could not be assumed that no count meant that the particle had spin -1, it could have had spin $+1$ and just been 'missed'. To take this into account it was necessary to run the experiment with one or both of the spin detectors removed, and then to use a modified form of the Bell inequality. We refer to the experimental papers, listed in the bibliography, for details.

The important quantity that is measured is a suitably normalised coincidence counting rate, which is predicted by quantum theory to be given by

$$R(a,b) = \tfrac{1}{4}[1 + 0.984 \cos 2(a-b)]. \qquad (5.10)$$

The factor 0.984, rather than unity, arises from imperfections in the detectors (some particles are missed). If this prediction holds throughout the whole range of angles then the Bell inequality is violated. In figure 26 we show the results. The agreement with quantum theory is perfect.

To demonstrate how effectively these results violate the Bell inequality, and hence forever rule out the possibility of a local realistic description of the world, the authors measured explicitly at the angles where the violation was maximum, namely with the configuration shown in figure 27, i.e. with $a - b = b - a' = a' - b' = 22.5°$, and $a - b' = 67.5°$. A particular quantity S which according to the Bell inequality has to be negative, but which according to quantum theory has to be 0.118 ± 0.005, is measured to be 0.126 ± 0.014. It is very clear that quantum theory and not locality wins.

In the next set of experiments both spin directions were explicitly detected, so the set-up was closer to that envisaged in the proof of the original Bell inequality. From the measurements, the value of $\langle F(a, a', b, b') \rangle$, defined in the previous section, was calculated as

$$F^{\text{expt}} = 2.697 \pm 0.015 \qquad (5.11)$$

for the orientation given by figure 27. This exceeds the bound given in the inequality by more than 40 times the uncertainty. On the other hand it agrees perfectly with the prediction of quantum theory which, again allowing for the finite size of the detectors, is calculated to be

$$F^{\text{qt}} = 2.70 \pm 0.05 \qquad (5.12)$$

instead of $2\sqrt{2}$, which is the result with perfect detectors.

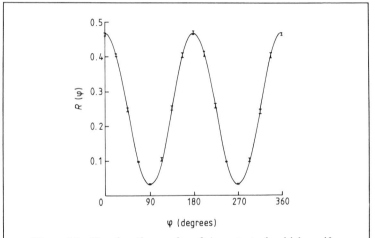

Figure 26   Showing the results of Aspect *et al*, which verify the predictions of quantum theory (the full curve) and so confirm a violation of the Bell inequality.

The third experiment was designed to investigate the following question. Quantum theory suggests that measurement at A, say, causes an instantaneous change at B, and this seems to be confirmed by experiment. It appears therefore that 'messages' are sent with infinite velocity (see the next section for further discussion of this). Such a requirement would, however, not be needed if it were assumed that the spin detecting instruments somehow communicate their orientations to each other prior to the emission of the photons, rather than when a photon actually reaches a detector. In order to eliminate this possibility it is necessary to arrange that the orientations are 'chosen' after the photons have been emitted. Clearly the time involved is too small to allow the rotation of mechanical measuring devices, so the experiment had two spin detectors at each side, with pre-set orientations, and used switching devices to deflect the photons into one or the other detector. The switches were independently controlled at random. Thus, when the photons were emitted, the orientations that were to be used had not been decided. We refer to the original paper for further details of this experiment and here record only the result, which was again in complete agreement with quantum theory, and in violation of the Bell inequality. Of course, it could be that nothing is really random and that the devices that controlled the switching themselves communicated with each other prior to the start of the experiment. Such bizarre possibilities are hard to rule out (though if we were sufficiently clever we could arrange that the signals which switch the detectors originate from distant, different, galaxies that, according to present ideas of the evolution of the universe, can never previously have been in any sort of communication).

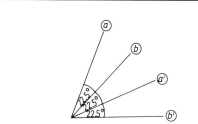

Figure 27 Showing the relative orientations of the spin measurements for maximal violation of the Bell inequality with spin one particles.

In this series of experiments it was also possible to vary the distance between the two detectors and so test whether the wave-function showed any sign of 'reducing' as a function of time, as it would according to the type of theory discussed in §3.4. Even when the separation was such that the time of travel of the photons was greater than the lifetime of the decaying states that produced them (which might conceivably be expected to be the time scale involved in such an effect), there was no evidence that this was happening.

Thus it appears that, once again, quantum theory has been gloriously successful. Maybe most of the people who regularly use it are not surprised by this; they have learned to live with its strange non-locality. The experiments we have described confirm this feature of the quantum world; no longer can we forget about it by pretending that it is simply a defect of our theoretical framework.

We close this section by noting the interesting irony in the history of the developments following the EPR paper. Einstein believed in reality (as we do); quantum theory seemed to deny such a belief and was therefore considered by Einstein to be incomplete. The EPR thought experiment was put forward as an argument, in which the idea of locality was implicitly used, to support this view. We now realise, however, that the experiment actually demonstrates the impossibility of there being a theory which is both complete and local.

5.6   Can signals travel faster than light?

According to the special theory of relativity, the velocity of light (or, more generally, of electromagnetic radiation) in vacuum is a fundamental property of time and space. The rules for combining velocities, and the laws of mechanics, etc, ensure that nothing can move with a velocity that exceeds this.

It would take us too far outside the scope of this book to explain special relativity; we can, however, assert with confidence that it is now firmly based on experimental observation and that it is a vital ingredient of the structure of contemporary theoretical physics. That its effects are not immediately obvious in our everyday experience is due to the large size of the velocity of light,

$$c = 3 \times 10^8 \text{ m s}^{-1}.$$

How then do we understand the fact that, according to quantum theory, wavefunction reduction happens instantaneously over arbitrarily large distances and, further, that such behaviour is apparently confirmed by experiment?

The first thing to notice here is that we cannot actually use this type of wavefunction reduction to transmit real messages from one macroscopic object to another. To help us appreciate what is meant by this statement we should distinguish the transmission of a message between two observers from what happens when the two observers both receive a message. For example, two people, one on Earth and one on Mars, could make an agreement that they will meet at a particular time either on Earth or on Mars. In order to determine which, they might agree to measure spins, in a pre-arranged direction, of electrons emitted in a particular EPR experiment. If they obtained $+1/2$ they would wait on their own planet, whereas if they obtained $-1/2$ they would travel to the other's planet. The correlation between the results of their measurements, noted in §5.4, would ensure that the meeting would take place. It would be possible for them to make their measurements at the same time, so they would receive the message telling them the place of the meeting simultaneously. However this message would not have been *sent* from one to the other.

We contrast this with the situation where the prior agreement is that the person on Earth will decide the venue and then try to communicate this to the person on Mars. How can he use the EPR type of experiment to transmit this message? The only option he has is either to make a measurement of the spin of the electron or not to make the measurement. A code could have been agreed: the measurement of the spin of A along a previously decided direction would mean that the meeting is to be on Earth, whereas no such measurement would mean that Mars would be the venue. Thus, at a particular time, he decides on his answer—he either makes the measurement or he does not. *Immediately* the wavefunction of B 'knows' this answer; in particular, if it is Earth then B will have a definite spin along the chosen direction, otherwise it will not.

The person on Mars, however, although he can observe the particle B, cannot 'read' this information because he is not able to measure a wavefunction. There is no procedure that the observer could use that would allow him to know whether or not the spin of B was definite or not.

The same conclusion is reached if we use, instead of a single experiment, an ensemble of identical experiments. In this case, if we decide on the venue Earth, then we would measure the spins of all the A particles in the specified direction. This would immediately mean that all the B particles had a definite spin in that direction. Now, *if these were all the same*, e.g. if they were all + 1/2, then we could verify this by simply measuring them. However, they would not all be the same, half would be + 1/2 and half would be − 1/2, which is exactly the same distribution we would have obtained if the spins were not definite, i.e. if the venue had been Mars and no measurements of A had been made.

The situation could be very different if the quantum theory description is incomplete and there are hidden variables. If these could, by some as yet unknown means, be measured, then, since measurements at A inevitably change these variables at B, the possibility of sending messages at an infinite velocity would seem to exist, in violation of the theory of special relativity. Such a violation can be seen explicitly in some types of hidden-variable theories where a quantum force is required to act instantaneously over arbitrarily large distances. This contrasts with the known forces, which in fact are due to exchange of particles and whose influence therefore cannot travel faster than the velocity of light.

We here have another very unpleasant feature of hidden-variable theories. It is not, however, possible to use this argument to rule them out entirely. Special relativity has only been tested in experiments that do not measure hidden variables; if we ever find ways of measuring them then the theory might be shown to be wrong— generalising results from one set of experiments to an entirely different set has often led to mistakes.

Even within normal quantum mechanics the question of how a wavefunction can reduce instantaneously, consistently with special relativity, is one that requires an answer. To discuss it would take us into relativistic quantum field theory, which is the method by which quantum theory and special relativity are combined. Although this theory has had many successes, it is certainly not fully understood and at the present time does not appear to have anything conclusive to say.

5.7 Summary of Chapter Five

We saw in earlier chapters that, according to quantum theory, a measurement on a system alters that system in a way which is indeterminate and is also inconsistent with the assumption that quantum theory can be applied to everything. If, however, reality is truly expressed in terms of hidden variables, then we can eliminate these features. There is no indeterminism in physics and no inexplicable wavefunction reduction, because measurements play the same role as in classical physics, i.e. they record what is already 'there'. The price we pay for this is that measurements must then affect other objects regardless of their distance from the object being measured. If somebody decides to measure something in Moscow then it will immediately change things here in Durham. This is an experimental fact. A local description of reality which permits us to talk of objects which are spatially isolated from each other does not exist.

We have here reached the ultimate 'silliness' of the quantum world. The lack of locality in quantum theory was seen in the very early days by Schrödinger as its prime feature: 'When two systems of which we know the states by their respective representations enter into a temporary physical interaction due to known forces between them and then, after a time of mutual influence, the systems separate again, then they can no longer be described in the same way as before viz. by endowing each of them with a representative state vector. I would not call that *one* but rather *the* characteristic of quantum mechanics.' We are now sure that this characteristic is not just a property of the quantum theoretical description of the world, rather it is property of the world itself. The only way it can be avoided is to adopt the many-worlds interpretation.

Non-local hidden-variable theories, that are either deterministic or not, do exist, and we have given a brief outline of some of their properties. Although they are in some respects very unnatural, they have the enormous advantage over all other theories of being precise and well defined. They require no arbitrary limitations of the theory to systems without observers.

Chapter Six

The Mysteries of the Quantum World

6.1 Where are we now?

Readers who have read this far are probably confused. Normally this is not a good situation to be in at the start of the last chapter of a book. Here, however, it could mean that we have at least learned something: the quantum world is very strange. Certain experimentally observed phenomena contradict any simple picture of an external reality. Although such phenomena are correctly predicted by quantum theory, this theory does not explain how they occur, nor does it resolve the contradictions.

What else ought we to have learned? We have seen, again on the basis of experiment, that a local picture of reality is false. In other words, the assumption that what happens in a given region of space is not affected by what happens in another, sufficiently distant, region is contrary to observation.

Nothing else is certain. We have met questions which appear to have several possible answers. None of these answers, however, are convincing. Indeed, it is probably closer to the truth to say that all are, to our minds, equally implausible. The quantum world teaches us that our present ways of thinking are inadequate.

I have tried to give a quick survey of the questions and their possible answers in tables 6.1 and 6.2 The first of these tables presents the problem purely in terms of the potential barrier experiment introduced in §1.3. No reference is made here to quantum theory or its concepts. The second table, on the other hand,

Table 6.1 Review of the problems associated with the potential barrier experiment.

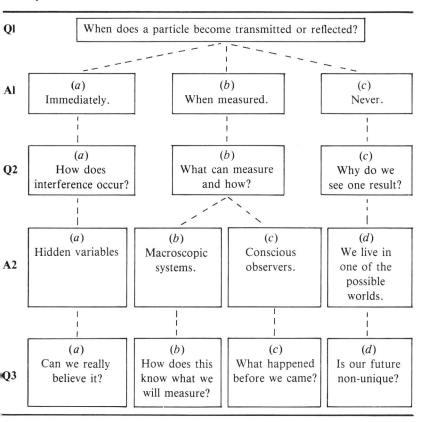

Notes

Q1: Here, 'when' could mean at what stage, or even how?

A1(*a*): 'Immediately' implies that the particle always goes one way or the other, regardless of whether it is seen.

A1(*b*): This means that the particle is not either transmitted or reflected until it reaches some 'measuring device'.

Q2(*a*): If the particle has followed one path how can it know about the mirror on the other side?

A2(*a*): Interference is not really happening. The trajectories of the particles make it appear that it is. The information about the other mirror comes through the quantum force which is non-local.

A2(*b*): Presumably when a system reaches a certain size or complexity it has the power to do this.

Q3: This is a sample of many possible questions at this stage.

Table 6.2 Review of the interpretation problems of quantum theory.

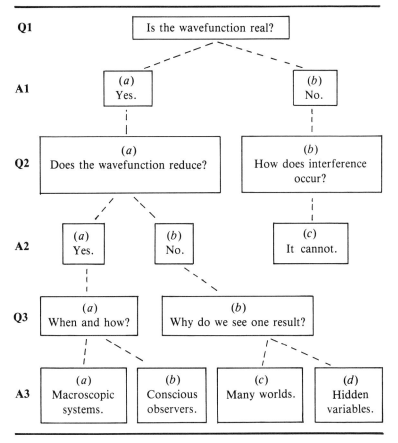

Notes

Q2(*a*): Or, does it change in some way in addition to the (deterministic) change with time implied by the Schrödinger equation?

Q2(*c*): If the wavefunction is merely our knowledge of the system, then it cannot give rise to interference. Nobody has found any way of obtaining the predictions of interference without using the wavefunction.

Q4: Many questions, similar to those in Q3 of table 6.1, could be asked here.

expresses the parallel questions from the point of view of quantum theory.

Readers should be very wary of regarding these tables as a substitute for the preceding text, since they are greatly over-simplified. The notes that are appended to each table are intended to clarify a few points.

These tables provide, in a very abbreviated form, a rough guide to our present understanding (or lack of understanding) of the quantum world. The four answers given in A3 of table 6.2 summarise what are, in my opinion, the available options for a solution to the interpretation problem of quantum theory. To some extent the first two differ only in degree so, if they are combined, we are left with three possibilities:

(*a*) and (*b*) Some modification to the rules of quantum theory which, for a sufficiently complex system, guarantee reduction of the wavefunction.

(*c*) The many-worlds interpretation.

(*d*) Hidden variables.

In the remainder of this chapter we shall briefly look at any other relevant facts, see what the possibilities are of further progress, present a few other opinions, and then conclude.

6.2 Quantum theory and relativity

This is a difficult section, from which we shall learn little that has obvious relevance to our theme. Nevertheless, the section must be included since its subject is very important and is an extremely successful part of theoretical physics. There is also the possibility, or the hope, that it could one day provide the answers to our problems.

The mysteries that we met in Chapter One arose from certain experimental facts. We have learned that quantum theory predicts the facts but does not explain the mysteries. Now we must learn that quantum theory also meets another separate problem, namely that it is not compatible with special relativity.

The reason for this is that special relativity requires that the laws of physics be the same for all observers regardless of their velocity (provided this is uniform). This requirement implies that only relative velocities are significant, or, in other words, that there is no meaning to absolute velocity. In practice this fact makes little difference to physics at low velocity; it is only when velocities become of the order of the velocity of light (3×10^8 m s^{-1}) that the new effects of special relativity are noticed.

Quantum theory, as originally developed, did not have this property of being independent of the velocity of the observer, and is thus inconsistent with special relativity. Although the practical effects of this inconsistency are very tiny for the experiments we have discussed, there are situations where they are important, and it is natural to ask whether quantum theory can be modified to take account of special relativity, and even to ask whether such modifications might provide some insight into our interpretation problems. The answer to the first of these questions is a qualified 'yes'; to the second it is a tentative 'no'.

The relativistic form of quantum mechanics is known as relativistic quantum field theory. It makes use of a procedure known as *second quantisation*. To appreciate what this means we recall that, in the transition from classical to quantum mechanics, variables like position changed from being definite to being uncertain, with a probability distribution given by a wavefunction, i.e. a (complex) number depending upon position. In relativistic quantum field theory we have a similar process taken one stage further: the wavefunctions are no longer definite but are uncertain, with a probability given by a 'wavefunctional'. This wavefunctional is again a (complex) number, but it depends upon the wavefunction, or, in the case where we wish to talk about several different types of particle, upon several wavefunctions, one for each type of particle. Thus we have the correspondence:

First quantisation: $x, y, \ldots$ replaced by $W(x, y, \ldots)$

Second quantisation: $W(x), U(x), \ldots$ replaced by $Z(W(x), U(x) \ldots)$.

The analogue of the Schrödinger equation now tells us how the wavefunctional changes with time.

An important practical aspect of relativistic quantum field theory is that the total number of particles of a given type is not a fixed number. Thus the theory permits creation and annihilation of particles to occur, in agreement with observation.

For further details of relativistic quantum field theory we must refer to other books. (Most of these are difficult and mathematical. An attempt to present some of the features in a simple way is made in my book *To Acknowledge the Wonder: The story of fundamental physics*, referred to in the bibliography.) There is no doubt that the theory has been enormously successful in explaining observed phenomena, and has indeed been a continuation of the success story of 'non-relativistic' quantum theory which we outlined in §2.5. In particular, it incorporates the extremely accurate predictions of quantum electrodynamics, has provided a partially unified theory of these interactions with the so-called weak interactions, and has provided us with a good theory of nuclear forces. In spite of these successes there are formal difficulties in the theory. Certain 'infinities' have to be removed and the only way of obtaining results is to use approximation methods, which, while they appear to work, are hard to justify with any degree of rigour.

Do we learn anything in all this which might help us with the nature of reality? Apparently not. If, in our previous, non-relativistic, discussion, we regarded the wavefunction as a part of reality, we now have to replace this by the wavefunctional, which is even further removed from the things we actually observe. The wavefunctions have become part of the observer-created world, i.e. things that become real only when measured.

We must now consider the problem of making quantum theory consistent with *general* relativity. Since general relativity is the theory of gravity, this problem is equivalent to that of constructing a quantum theory of gravity. Much effort has been devoted to this end, but a satisfactory solution does not yet exist. Maybe the lack of success achieved so far suggests that something is wrong with quantum theory at this level and that, if we knew how to put it right, we would have some clues to help with our interpretation problem. This is perhaps a wildly optimistic hope but there are a few positive indications. Gravity is negligible for small objects, i.e. those for which quantum interference has been tested, but it might become important for macroscopic objects, where, it appears, wavefunction reduction occurs. Could gravity somehow be the small effect responsible for wavefunction reduction, as discussed in §3.7?

Probably the correct answer is that it cannot, but if we want encouragement to pursue the idea we could note that the magnitudes involved are about right. The ratio of the electric force (which is responsible for the effects seen in macroscopic laboratory physics) to the gravitational force, between two protons, is about 10^{36}. For larger objects the gravitational force increases (in fact it is proportional to the product of the two masses), whereas this tends not to happen with the electric force because most objects are approximately electrically neutral, with the positive charge on protons being cancelled by the negative charge on electrons. Consider, then, the forces between two massive objects, each of which has charge equal to the charge on a proton. The electric force will be equal to the gravitational charge if the objects weigh about 10^{-6} g. Thus we can see that gravitational forces become of the same order as electrical forces only when the objects are enormously bigger than the particles used in interference effects, but that they are certainly of the same order by the time we reach genuine macroscopic objects. (See also the remarks at the end of Appendix 7.)

We end this section by noting a few other points. General relativity is all about time and space, about the fact that our apparently 'flat' space is only an approximation, about the possibility that there are singular times of creation, and/or extinction, about the existence of black holes with their strange effects. Some of these facts could be relevant, but at the present time all must be speculation. As an example of such speculation we mention the suggestion of Penrose that there might be some sort of trade-off between the creation of black holes and the reduction of wave packets (see the acticle by Penrose, 'Gravity and State Vector Reduction' in *Quantum Concepts in Space and Time*, ed C J Isham and R Penrose [Oxford: Oxford University Press 1985]).

6.3   Where do we go next?

Quantum theory has been the basis of almost all the theoretical physics of this century. It has progressed steadily, indeed gloriously. The early years established the idea of quanta, particularly for light, then came the applications to electrons which led to all the developments in atomic physics and to the solution of chemistry, so that already in 1929 Dirac could write that 'The underlying physical laws necessary for the mathematical theory of a large part of physics and the whole of chemistry are thus completely known...' (*Proceedings of the Royal Society* A 123 714). The struggle to combine quantum theory with special relativity, discussed in the preceding section, occupied the period from the 1930s to the present, and its successes have ranged from quantum electrodynamics to QCD, the theory of strong interactions. We are now at the stage where much is understood and there is confidence to tackle the remaining problems, like that of producing a quantum theory of gravity.

The interpretation problem has been known since the earliest days of the subject (recall Einstein's remark mentioned in §1.1), but here progress has been less rapid. The 'Copenhagen' interpretation, discussed in the next section, convinced many people that the problems were either solved or else were insoluble. The first really new development came in 1935 with the EPR paper, which, as we have seen, purported to show that quantum theory was incomplete. We must then wait until the 1950s for Bell's demolition of the von Neumann argument regarding the impossibility of hidden-variable theories, and, later, for his theorem about possible results of local theories in the EPR experiment. Throughout the whole period there were also steady developments leading to satisfactory hidden-variable theories. At present, attempts are being made to see if these are, or if they can be made, compatible with the requirements of special relativity.

What progress can we expect in the future? In the very nature of the case, new insights and exciting developments are unlikely to be predictable. We can, however, suggest a few areas where they might occur.

Let us consider, first, possible experiments. There is much interest at present in checking the accuracy of simple predictions of quantum theory, in order, for example, to see whether there is any indication of non-linear effects. No such indications have been seen at the present time, but continuing checks, to better accuracy and in different circumstances, will continue to be made.

Another area where there is active work being done is in the possibility of measuring interference effects with macroscopic objects, or at least with objects that have many more degrees of freedom than electrons or photons. The best hope for progress here lies in the use of SQUIDs (superconducting quantum interference devices). These are superconducting rings, with radii of several centimetres, in which it is hoped that interference phenomena, as predicted by quantum theory, between currents in the rings can be observed. Such observations will verify (or otherwise) the predictions of quantum theory for genuinely macroscopic objects. In particular, it should be possible to see interference between states that are *macroscopically* different, and thereby verify that a system can be in a quantum mechanical superposition of two such states (cf the discussion of Schrödinger's cat, etc, in §4.3).

The success of quantum theory, combined with its interpretation problems, should always provide an incentive to experimentalists to find some result which it cannot predict. Many people would probably say that they are unlikely to find such a result, but the rewards for so doing would be great. If something could be shown to be wrong with the experimental predictions of orthodox quantum theory then we would, at last, perhaps have a real clue to understanding it.

It must be admitted that the likelihood of there being any practical applications arising from possible discoveries in this area is extremely low. There are many precedents, however, that should prevent us from totally excluding them. We have already noted in §5.6 that genuine observation of wavefunctions, were it ever to be possible, might lead to the possibility of instantaneous transmission of signals. To allow ourselves an even more bizarre (some would say ridiculous) speculation, we recall that, as long as the wavefunction is not reduced, then all parts of it evolve with time according to the Schrödinger equation. Thus, for example, the quantum world contains the complete story of what happens at all subsequent times to both the transmitted and reflected parts of the wavefunction in a barrier experiment. Suppose then that a computer is programmed by a non-reduced wavefunction which contains many different programs. In principle this is possible; different input keys could be pressed according to the results ('unobserved', of course) of a selection of barrier type experiments, or, more easily, according to the spin projections of particles along some axis. As long as the wavefunction is not reduced, the computer performs all the programs simultaneously. This is the ultimate in parallel processing! If we observe the output answer by normal means we select one set of results of the experiments, and hence one program giving a single answer. The unreduced output wavefunction, however, contains the answers to *all* the programs. It is unlikely that we will ever be able to read this information, but ...

On the theoretical side, we have already mentioned the possibility that the difficulties with making a quantum theory of gravity just might be related to the defects of quantum theory. Maybe some of our difficulties with non-locality suggest that our notions of time and space are incomplete. If, for example, our three dimensions of space are really embedded in a space of more dimensions then we might imagine that points of space which seem to us to be far separated are in reality close together (just as the points on a ball of string are all close, except to an observer who, for some reason, can only travel along the string).

Bearing in mind the issue of causality, we might ask why we expect this to exist in the first place, in particular, why we believe that the past causes the present. Indeed we could wonder why there is such a difference between the past, which we remember, and the future, which we don't! In case we are tempted to think these things are just obvious, we should note that the fundamental laws of physics are completely neutral with regard to the direction of time, i.e. they are unchanged if we change the sign of the time variable. In this respect time is just like a space variable, for which it is clear that one direction is not in any fundamental respect different from any other. Concepts like 'past' and 'present', separated by a 'now', do not have a natural place in the laws of physics. Presumably this is why Einstein was able to write to a friend that the distinction between past and present was only a 'stubbornly persistent illusion'.

It may well be that, in order to understand quantum theory, we need totally new ways of thinking, ways that somehow go beyond these illusions. Whether we will find them, or whether we are so conditioned that they are for ever outside our scope is not at present decidable.

6.4 Early history and the Copenhagen interpretation

We have not, in this book, been greatly concerned with the historical development of quantum theory. When an idea is new many mistakes are made, blind alleys followed, and the really significant features can sometimes be missed. Thus history is unlikely to be a good teacher. Nevertheless, it is of interest to look back briefly on how the people who introduced quantum theory into physics interpreted what they were doing.

Already we have noted that Einstein, surely the premier scientist of this century, was always unhappy with quantum theory, which he considered to be, in some way, incomplete. Initially his objections seemed to be to the lack of causality implied by the theory, and to the restrictions imposed by the uncertainty principle. He had a long running controversy with Bohr on these issues, a controversy which it is fair to say he lost. In addition, however, Einstein was one of the first to realise the deeper conceptual problems. These he was not able to resolve. Many years after the time when he was the first to teach the world about photons, the particles of light, he admitted that he still did not understand what they were.

Even more remarkable, perhaps, was the attitude of Schrödinger. We recall that it was he who introduced the equation that bears his name, and which is *the* practical expression of quantum theory, with solutions that contain a large proportion of all science. In 1926, while on a visit to Copenhagen for discussions with Bohr and Heisenberg, he remarked: 'If all this damned quantum jumping were really to stay, I should be sorry I ever got involved with quantum theory.' (This quote, which is of course a translation from the original German, is taken from the book by Jammer, *The Philosophy of Quantum Mechanics*, p 57). The 'jumping' presumably refers to wavefunction reduction, a phenomenon Schrödinger realised was unexplained within the theory, which he, like Einstein, therefore regarded as incomplete. To illustrate the problem in a picturesque way he invented, in 1935, the 'Schrödinger cat' story, which we have already discussed in §4.4. He considered it naive to believe that the cat was in an uncertain, dead or alive, state until observed by a conscious observer, and therefore concluded that the quantum theory could not be a proper description of reality.

Next we mention de Broglie, who, it will be recalled, was the first to suggest a wave nature for electrons. He was also unhappy with the way quantum theory developed, and took the attitude that it was wrong to abandon the classical idea that particles followed trajectories. He believed that the role of the wavefunction was to act as a pilot wave to guide these trajectories, an idea which paved the way for hidden-variable theories.

Thus, of the four people (Planck, Einstein, Schrödinger, de Broglie) who probably played the leading roles in starting quantum theory, three became, and remained, dissatisfied with the way it developed and with its accepted 'orthodoxy'. This orthodoxy is primarily due to the other three major figures in the early development of the theory, Bohr and, to a lesser extent, Heisenberg and Born. It has become known as the 'Copenhagen' interpretation.

A precise account of what the Copenhagen interpretation actually is does not exist. Quotations from Bohr's articles do not always seem to be consistent (which is not surprising in view of the fact that the ideas were being developed as the articles were being written). Almost certainly, two present-day physicists, who both believe that they subscribe to the orthodox (Copenhagen) interpretation, would give different accounts of what it actually means. Nevertheless there are several key features which, with varying degrees of emphasis, would be likely to be present. We shall endeavour to describe these.

(i) Bohr made much use of the notion of 'complementarity': particle and wave descriptions complement each other; one is suitable for one set of experiments, the other for different experiments. Thus, since the two descriptions are relevant to different experiments, it does not make sense to ask whether they are consistent with each other. Neither should be used outside its own domain of applicability.

(ii) The interpretation problems of quantum theory rest on classical ways of thinking which are wrong and should be abandoned. If we abandon them then we will have no problems. Thus questions which can only be asked using classical concepts are not permitted. Classical physics enters only through the so-called 'correspondence' principle, which says that the results of quantum theory must agree with those of classical mechanics in the region of the parameters where classical mechanics is expected to work. This idea, originally used by Planck, played an important role in the discovery of the correct form of quantum theory.

(iii) The underlying philosophy was strongly 'anti-realist' in tone. To Bohr: 'There is no quantum world. There is only an abstract quantum physical description. It is wrong to think that the task of physics is to find out how nature *is*. Physics concerns what we can say about nature.' Thus the Copenhagen interpretation and the prevailing fashion in philosophy, which inclined to logical positivism, were mutually supportive. The only things that we are allowed to discuss are the results of experiments. We are not allowed to ask, for example, which way a particle goes in the interference experiment of §1.4. The only way to make this a sensible question would be to consider *measuring* the route taken by the particle. This would give us a different experiment for which there would not be any interference. Similarly, Bohr's reply to the alleged demonstration of the incompleteness of quantum theory, based on the EPR experiment, was that it was meaningless to speak of the state of the two particles prior to their being measured. (It should be noted that Einstein himself had made remarks which were in this spirit. Indeed Heisenberg, a convinced advocate of the Copenhagen interpretation, was apparently helped along this line by one such remark: 'It is the theory which decides what we can observe.')

(iv) All this leaves aside the question of what constitutes a 'measurement' or an 'observation'. It is possible that somewhere in the back of everyone's mind there lurked the idea of apparatuses that were 'classical', i.e. that did not obey the rules of quantum theory. In the early days the universality of quantum theory was not appreciated, so it was more reasonable to divide the world into, on the one hand, observed systems which obeyed the rules of quantum mechanics, and, on the other, measuring devices, which were classical.

These, then, are the ingredients of the Copenhagen interpretation. It is very vague and answers few of the questions; anybody who thinks about the subject today would be unlikely to find it satisfactory: yet it became the accepted orthodoxy. We have already, in §5.2, suggested reasons why this should be so. The theory was a glorious success, nobody had any better answers to the questions, so all relaxed in the comfortable glow of the fact that Bohr had either answered them or told us that they should not be asked.

I was a research student in Manchester in the 1950s. Rosenfeld was the head of the department and the Copenhagen interpretation reigned unquestioned. One particular Christmas, the department visited the theoretical physics department in Birmingham to sing carols (that, at least, was the excuse). Some of the carols were parodied. In particular, I remember the words we used for the carol that normally begins 'The boar's head in hand bear I'. They were:

At Bohr's feet I lay me down,
For I have no theories of my own
His principles perplex my mind,
But he is oh so very kind.
Correspondence is my cry, I don't know why,
  I don't know why.

But we were all afraid to ask!

6.5 A bibliography and some recent opinions

In this section we shall suggest a few books to which interested readers should turn for more information on some of the topics we have discussed. We shall also note the conclusions of recent authors on the controversial issues.

There is an excellent selection of textbooks available for those who wish to know more about the details of quantum mechanics as a straightforward physical theory, and to see how it is applied to various real problems. Such books require their readers to be familiar with a certain amount of advanced mathematics. Here we mention only four, namely Schiff, *Quantum Mechanics* [New York: McGraw-Hill 1949] (because it was the book from which I first learned the subject), Dirac, *The Principles of Quantum Mechanics* [Oxford: Oxford University Press 1930] (because it demonstrates the elegance of formal quantum theory), Gottfried, *Quantum Mechanics* [New York: Benjamin 1965] (because it contains a careful discussion of the measurement process and the difference between mixtures and pure states) and, finally, Cohen-Tannoudji, Diu and Laloë, *Méchanique Quantique* [Paris: Herman 1973] (because it is modern, well written and complete). There are, however, many others probably equally good.

For reasonably simple, 'popular', accounts of the recent progress that has been made in understanding the micro-world of elementary particles, using quantum theory, there are several books available, e.g. Close, *The Cosmic Onion* [London: Heinemann 1983], and my own book, *To Acknowledge the Wonder: The Story of Fundamental Physics* [Bristol: Adam Hilger 1985]. A thorough discussion of the anthropic principle can be found in Paul Davies' book, *The Accidental Universe* [Cambridge: Cambridge University Press 1982] and also in the more recent book by Barrow and Tipler, *The Anthropic Cosmological Principle* [Oxford: Oxford University Press 1985].

An excellent, though technical, account of the EPR experiments, and of Bell's theorem in its various forms, is given in a review by Clauser and Shimony (*Reports on Progress in Physics* **41** 881 (1978), reprinted in *Lasers in Applied and Fundamental Research* ed S Stenholm [Bristol: Adam Hilger 1985]). Further details of the Aspect *et al* experiments can be found in the original papers published in *Physical Review Letters* **47** 460 (1981) and **49** 91 and 1804 (1982).

Turning now to the interpretation problems of quantum theory, four recent books, all written for non-specialist readers, must be noted: *The Quantum World* by Polkinghorne [London: Longman 1984], *The Cosmic Code* by Pagels [London: Michael Joseph 1982], *In Search of Reality* by d'Espagnat [Berlin: Springer 1983] and *In Search of Schrödinger's Cat* by Gribbin [London: Wildwood 1984].

Polkinghorne clearly does not accept hidden-variable theories (which he likens, perhaps a little unfairly, to the epicycle explanation for planetary orbits), and is equally unhappy with any suggestion that consciousness might play any role. He is also a realist, dismissing the idea that a retreat into positivism provides a satisfactory solution. This, he admits, leads him to no conclusion except that: 'It is a curious tale.' He makes an interesting comparison between the difficulty of passing from the micro-world to the macro-world, and similar difficulties when we try to pass from, for example, the world of physics to that of biology or of life. Although each 'level' is in some way a consequence of what happens at a deeper, more reductionist, level, there remains a degree of 'level autonomy'.

Pagels gives a picturesque account of a variety of shops that endeavour to sell their own interpretation of quantum theory. He does not recommend that we buy from any, but, rather, suggests that they all contain some aspect of the truth. He draws an analogy with a picture that has the property of seeming to represent one thing when we see the dark part, but which appears to be another when we see the light part. This book also gives a very readable account of other aspects of modern physics.

D'Espagnat's book is more philosophic in tone than either of the above. It concentrates very much on the implications of the Aspect *et al* verifications of the quantum theoretical predictions of the EPR type of experiment. There is a clever classical analogue which beautifully demonstrates the amazing non-locality of nature which these results prove. Although, like everyone else, d'Espagnat cannot fully comprehend the message of the quantum world, he is convinced of its importance: 'when the statement has been made that physics accounts for almost all phenomena, the main contribution of this science to basic knowledge has *not yet* been formulated... the truly basic contribution of contemporary physics is essentially contained in the dichotomy that this science ...seems to establish between Being and objects or between reality and phenomena.' He believes in a reality, but emphasises that the elements of that reality are not the notions of everyday life: it is, in his words, a 'far reality'.

Gribbin's book, perhaps because it is written at a more elementary level than the others, is pleasantly easy to read. There is some oversimplification and certain topics are treated very briefly or omitted entirely. The author strongly supports the many-worlds interpretation.

Further discussion of the many-worlds interpretation can be found in the book *The Many Worlds Interpretation of Quantum Mechanics* ed DeWitt and Graham [Princeton: Princeton University Press 1973]. A good review of the model, in particular of its possible relevance to cosmology, is also given by Tipler in his article 'The Many Worlds Interpretation of Quantum Mechanics in Quantum Cosmology' published in *Quantum Concepts in Space and Time* ed Isham and Penrose [Oxford: Oxford University Press 1985].

Next, I must mention the recent book by Wallace Garden, *Modern Logic and Quantum Mechanics* [Bristol: Adam Hilger 1983]. This is very different from the previous books because it uses the language and methods of formal logic to discuss quantum theory. The author describes the ideas of so-called quantum logic, a structure which is designed to accord with the results of quantum theory (it uses *yes, no* and *undecidable* to replace the usual two-valued logic). She gives reasons for believing that this does not solve any of the problems, and presents an alternative which leaves her with the conclusion: 'This logical interpretation of quantum-mechanics can claim a measure of sense and reason. It is surely sensible to see the quantum peculiarities as products of weak description rather than as incomprehensible features of the world. It is surely more reasonable to suppose that our theory is inadequate than to argue that reality itself is bizarre. . . We should accept quantum mechanics as the most successful theory we have at present, while setting out to develop a new and better theory of reality.'

For a good account of the early history of quantum theory, and in particular of the contribution of Einstein, readers should consult the biography of Einstein, *Subtle is the Lord*, by Pais [Oxford: Oxford University Press 1982]. The best attempt to explain the Copenhagen interpretation that I have seen is the article of Stapp published in the *American Journal of Physics* **40** 1098 (1972). This article includes some interesting comments of Heisenberg.

The most complete descriptions of the history of quantum theory and its philosophical consequences are in the two books by Jammer, *The Conceptual Development of Quantum Mechanics* [New York: McGraw-Hill 1966] and *The Philosophy of Quantum Mechanics* [New York: Wiley 1974]. These are extremely thorough, and seem to contain everything, including many intriguing quotations. Jammer does not give us his own conclusions, taking instead the wiser course of ending the second of the above books with the following quotation: 'It is better to debate a question without settling it than to settle a question without debating it' (J Joubert).

Finally, we note that this brief review of some of the current literature is fully consistent with the remark of Feynman (one of the leading contributors to the development of relativistic quantum field theory) that 'nobody really understands quantum theory'.

6.6   A final plea for reality

In §1.2 we stated our reasons for believing in an external reality. Throughout this book we have endeavoured to use the clues provided by observation in order to understand the nature of that reality. These clues have led in directions that are not compatible and no convincing picture of the reality has emerged. Does this mean that the search was in vain? Are quantum phenomena telling us that the confidence we expressed in §1.2 was misplaced and that, in fact, there is no external reality, no truth that lies behind our observations? Are we looking for a pot of gold at the end of a rainbow?

As we have already noted, there are those who would certainly answer *yes* to these questions. They would claim that observation is all that there is and that the idea of external reality is simply an illusion. Such views dominated the thinking of philosophers in the years that followed the advent of quantum theory and, although their influence has probably declined and there have been many modifications in the details, they still appear to be widely held.

In order to be fair to the 'anti-realist' case, I shall state it in a modern form by quoting from the excellent book, *The Scientific Image* [Oxford: Oxford University Press 1980] by van Fraassen. He first defines what he calls the doctrine of 'scientific realism': 'Science aims to give us, in its theories, a literally true story of what the world is like; and acceptance of a scientific theory involves the belief that it is true.' This is a doctrine to which I would assent, and it has been the basis of the discussions in this book. It is consistent with a desire to explain observed phenomena and to understand the nature of what exists. (I would worry a little about the meaning of the word 'literally' in this context, but certainly I believe that the statement is true in spirit.)

Van Fraassen, however, rejects this doctrine and replaces it by what he refers to as the doctrine of 'constructive empiricism': 'Science aims to give us theories which are empirically adequate; and acceptance of a theory involves as belief only that it is empirically adequate.' The expression 'empirically adequate' means in agreement with all observations.

This doctrine is anti-realist. It implies that we should not seek to explain things, but should be satisfied with 'theories' that give results in agreement with observations. Clearly the Copenhagen interpretation of quantum theory is in accordance with this doctrine.

We are now entering an area in which there is a vast literature and which we cannot hope to explore fully here. The book by van Fraassen, referred to above, gives a very good account of the arguments in favour of the anti-realist view. It also contains many further references.

In answering the case against realism, as put by van Fraassen, I would note, first, that what constitutes science could perhaps be seen as simply a definition of the word 'science'. Then the two statements above would simply correspond to different definitions, and the issue would not have any great interest to physics. More reasonably, we could regard 'science' as something that scientists do, or that they believe they do. The issue then is a sociological one. I am not aware of the existence of any surveys, but I would imagine that most scientists would hold to the realist doctrine. To take a simple example of this, a few years ago I attended a party at CERN to celebrate the discovery of the W boson. There would have been no excuse to hold the party if scientists had not been convinced they had shown that these particles, predicted by theorists, really existed. Science is concerned with *discoveries*, not *inventions*!

Now, it is clear that many professional philosophers believe that scientists are mistaken in what they think they are doing. Leaving aside for the moment the obvious remark that people who are doing something might be better able to say what they are doing than people who are not, it is interesting to ask why this is so. Certainly one reason for the confidence that scientists have that they are discovering truth, comes from the tremendous success that science has had throughout, say, the last hundred years. Within physics there is justified confidence that we have really understood an enormous amount of the observed world, we have reduced a bewildering variety of at first sight chaotic phenomena to a few simple laws. It is perhaps worthwhile listing a few examples:

Newton's explanation of the properties of the planetary orbits in terms of his law of gravity.

The understanding of the properties of atoms which once were thought to be totally outside physics.

The understanding of the nuclear forces in terms of QCD.

The standard model of weak interactions, culminating in the successful prediction of the existence and masses of the W and Z particles.

The prediction of the microwave background from the properties of the universe before life, or even galaxies, were formed.

We could add to the list, but these will suffice for our purpose. Notice that we believe the theories are 'true', i.e. that atoms really are made of nuclei and electrons, that there really was a time when the universe was hot and in thermal equilibrium, etc. (On the other hand, the fantastic prediction of the existence and wavelength of the microwave background would not be regarded as a triumph for theoretical physics if we had some reasons for knowing that the appropriate conditions, given in the Big Bang model of the universe, never actually occurred.)

These successes have made physicists very confident. Many questions which once we would not have thought of asking, e.g. why does the world have three space dimensions? why does the electric charge have the value it does? why is the rate of expansion of the universe so close to the so-called critical value? etc, we now ask and even try to answer.

We thus have a very coherent and satisfactory understanding of most of the physical world (I am ignoring here, of course, the problems of quantum theory). On the other hand, it is hard for those outside physics to appreciate fully how successful the picture actually is. The seemingly endless sequence of new objects, with names that sometimes sound rather frivolous, can so easily be dismissed as the fantasy of physicists, particularly by those who may be a little bit jealous of the supposed successes. The simplicity, i.e. the fact that there is a standard model that explains all of observed physics in terms of a few simple ideas and a few parameters, is not easily grasped.

As an illustration of the sort of conflict between the view of a physicist and that of a philosopher I note that van Fraassen, in the book mentioned above, is clearly unhappy to believe that even electrons *exist*, through he accepts, for example, the moons of Jupiter. To a physicist, the evidence in both cases is of a similar nature. All such evidence comes to us ultimately through our senses, but there is always a chain of events between the objects we are observing and the actual sensation. In some cases this is very simple and obvious, in others it is a long chain and we accept the evidence only because we believe we have an understanding of the whole observation process. Where the methodology is new, and we have little corroborative evidence, then we are less certain (the recent observation of 'top' quarks is a good example), but the issues involved are then issues of physics, not philosophy. In cases where there are doubts, we will settle them, one way or the other, by doing better physics. Basically the reasons why I believe that electrons, or even quarks, exist are not different from those which persuade me to believe in the existence of the piece of paper on which I am writing. Another example of this unwillingness to accept the basic continuity in our description of nature is van Fraassen's confident assertion that whereas electrons interfere, bullets do not. According to quantum mechanics they interfere in *exactly* the same way (i.e. we could calculate the interference pattern from exactly the same basic equations, with of course different masses, sizes, etc). Whether quantum theory is *correct* for bullets is another matter; it is something about which, as we have noted, we would dearly love to know more.

Maybe then we can see some of the reasons why there is a difference between the attitudes of physicists and philosophers to what the former are doing. It is a difference that explains why van Fraassen says that quantum theory proves that 'explanation' is impossible, whereas most physicists would say that it encourages us to look harder for explanations.

But who is 'right'? (Since we are committed to a belief in 'truth' we cannot avoid this question.) No *certain* answer can be given, any more than we can be *certain* of the existence of anything. Nevertheless, I am tempted to say that the anti-realist case is merely an intellectual game, and that nobody actually believes in it (at least, as Polkinghorne says, 'outside of his study'). The anti-realists are forced to insist on a distinction between familiar objects, and things like electrons, etc, because they know that their arguments, applied to familiar things, are foolish. The statement, typical of anti-realist thinking, that no phenomenon is a phenomenon unlesss it is an observed phenomenon, cannot seriously be maintained. It seems to imply that the universe began when Adam opened his eyes, or that 'history' is books in the library and not a story of things that actually happened. The fact that historians might sometimes (or even often) have got it wrong is, of course, counter to the anti-realists' case, since without realism nothing actually happened, so there is no truth. Similarly the statement that two theories, both of which fit the data, are equally good can be seen to be unreasonable if we note that a theory in which the sun always turns into cream cheese as it disappears over the horizon, and turns back again later, gives a perfectly adequate account of my observations.

This, however, is probably being unfair, and it is certainly an oversimplification to represent a variety of different views and nuances as though they are identical, so readers should turn to the literature for further details and for arguments in support of the anti-realist position.

Although the various views appear to be in conflict, I sometimes think that, when they are carefully stated, the differences are quite small. Reconciliation may lie in the recognition that 'certainty' is an illusion, we can never be absolutely sure about anything. What we (I) believe to be true is that which accords most simply with the presently available data. Provided that in this I include my sense of what is reasonable, i.e. the 'data' in my brain, then maybe empirical adequacy is a complete criterion for what I believe to be true. To put essentially the same idea in another form, it could be that existence *is* consistency with the evidence. Or, in the words of the philosopher C S Peirce, 'Reality, then, is persistence, is regularity. In the original chaos, where there was no regularity, there was no existence.'

I will close this section, then, by giving what seems to me to be the pragmatic case for realism, i.e. for the belief in an external reality, or in other words for taking the first of the definitions of science given earlier. It is the belief that a reality exists that provides the motivation for seeking it. Built into our nature is our desire to know the truth; it is, along with beauty and love, one of the things for which we crave. Though we will surely never find it, in any absolute sense, we will go on seeking it. Certainly we have far better chances of finding it if we search for it, than if we accept that it doesn't exist and stop looking. Indeed, one can even go further and say that the fact that secretly we know we will never ultimately know the full truth is a great comfort, since a solved problem is a problem with no more interest. As long as there are questions that we can ask we shall go on asking them, and the fact that they are hard questions will not be a deterrent; rather it will provide the inspiration to work harder on finding their answers.

6.7 A last look at the wavefunction

Before thir eyes in sudden view appear
The secrets...
Without dimension, where length, bredth, and highth
And time and place are lost...†

† Milton, *Paradise Lost*, Book II.

In §3.4 we gave some arguments for believing that wavefunctions are part of the external reality, or rather that *the* wavefunction (which in principle should describe the whole world) is a part of this reality. We also saw that this belief met with some problems. The time has come to review the position in the light of what we have learned.

The main point to note is that all the solutions to the interpretation problem of quantum theory that we have met, and that we summarised in table 6.2, require the wavefunction to exist. Even hidden-variable theories do not eliminate this requirement. It seems, therefore, that we must accept that wavefunctions are real objects, and try to understand what they are and what is implied by their existence.

To begin at the beginning, our first primitive notion of existence refers to objects which have a particular location, i.e. they *are somewhere*. Normally, in fact, objects have a finite size, so they occupy a well defined region of space. This region can, of course, change with time, corresponding to the object moving. It is not hard to extend this idea of an object at a particular place to the idea of a 'density', e.g. of a gas. This density is not *at* a particular place, but rather it has a value at each point of space. This value represents the amount of the gas, or whatever, at the point considered. (More precisely, the amount in a small region around the point.) Again we do not necessarily have a static situation, so these values can vary with time.

Now, in a world of only one particle, the wavefunction is similar to a density; it has a (complex) value at each point in the region of space considered. We can therefore easily picture the wavefunction as being part of external reality. We can think of it as being 'something' spread out over space, with a particular amount at each point.

As we have seen already in §3.4, such a simple picture is not possible when we have more than one particle. If, for example, we consider a world containing two particles, which we denote by A and B, then the wavefunction associated with this world should tell me the probability of finding particle A at position x_A, and particle B at position x_B. Thus it would be a function of *two* positions which I might write as $W(x_A, x_B)$. Such a wavefunction is not like a density; it does not have a value at each point of space. To help us to appreciate this distinction we note that there are in fact some cases where the two-particle wavefunction can be written as a product of two wavefunctions, each depending upon only one position, i.e.

$$W(x_A, x_B) = U(x_A)V(x_B). \qquad (6.1)$$

When this occurs then W is represented by two functions which we can picture; there are now two 'somethings' which have densities at all points of space. However, it is a consequence of this product form for the wavefunction that the probability of finding one particle at a particular position is independent of the position of the other particle. This would only be true if the particles moved independently of each other, which is never the case in the real world. Indeed, even if the particles could be assumed to be not interacting, then the requirement of symmetry or antisymmetry §3.4 and Appendix 4) would ensure that W was not a simple product.

This problem naturally becomes much worse if we consider the true world of many particles, or more realistically the situation in quantum field theory, where W is dependent not upon *positions*, but upon *fields*. This is unlike anything we have previously thought of as real. We are therefore forced to conclude that we must enlarge our notion of what it means for something to exist. The wavefunction does not have a position and it does not make sense to ask how much of it there is in a particular region of space. The whole idea that things that *are*, are at particular points of space seems to be contradicted by quantum theory. Maybe this is the key thing that we have learned.

It is a hard lesson. Deeply embedded in our minds is the conviction that things that exist do so at particular places. But is it necessary? Is it a truth about reality, or just a limitation of our way of thinking? Quantum theory suggests the latter. It supports a remark made by Einstein (in a different context) that 'time and space are modes by which we think, not conditions in which we live'. Of course, in any really complete theory of the physical world we might want even the properties of space and time, even the number of spatial dimensions, to be calculable from the theory, so perhaps in some sense it is true to say that space itself is contained in the wavefunction rather than the other way around.

We can now see clear links with these ideas and the 'holistic' discussion in §4.4. Our prejudices are always to divide things into the very small, to try to describe everything in terms of what happens in small regions of space. Maybe there are things (indeed quantum theory says that the wavefunction *is* such a thing) that cannot be so subdivided.

There are also possible links here with the ideas of §3.7, where we suggested that non-linear modifications to quantum theory might be important in wavefunction reduction. To appreciate this we recall that in a linear theory the effect of two 'causes' is the sum of the effect of either cause separately, i.e. the two together yield both of the individual effects, but nothing new. Thus the effect of the 'whole' is the sum of the parts. Where there are important non-linear effects, this is not true, so dividing things into small, localised pieces loses something of the truth.

All this is very tentative and, at the present time, we cannot take it any further. It is worth noting however that, outside the world of physics, there is nothing surprising in what we are saying here. It is not normally expected that a thought should have a spatial location, or that it is readily divisible into pieces. Indeed, we have already seen that it is very hard to find any location for consciousness, and attempts to understand consciousness in terms of simpler, 'smaller', things do not seem to lead anywhere.

The realm of theology offers even better examples. Although 'primitive' (?) religions might place God at a particular place, such a way of thinking would not be normal to contemporary theologians. Adam and Eve apparently thought it possible to hide from God, Jonah even took a cruise to get away from him, but already the psalmist knew that

> *If I climb up to Heaven, Thou art there;*
> *if I make my bed in Sheol, again I find Thee.*
> *If I take my flight to the frontiers of the morning*
> *or dwell at the limit of the western sea,*
> *even there Thy hand will meet me*
> *and Thy right hand will hold me fast.*†

Theology has had to escape from a too restricted view of the reality of space; maybe physics is being called to follow.

---

† *Ps.* 139: 8–10.

6.8 Conclusions

God moves in a mysterious way, His wonders to perform†

We are conscious beings, able to receive through our senses information about the reality in which we live and of which we are a part. Through instruments which we have constructed we are able to probe deeply into this reality. Although we read these instruments at 'one end', through the medium of our senses, we believe that they give us information about details of a real world which exists at the 'other end' of the instrument. This belief is acquired partly as an extension of experience and partly because we have an understanding of how the instrument works.

In this way we have acquired the results of many observations of the external world. These results, however, are not the complete extent of reality; rather they are the clues from which we try to deduce what is, what was and even what will be. Thus, history is not merely records stored in the library, it is what happened to real people who led real lives and participated in real events. The investigations of geologists tell a story of the real evolution of the rocks upon our planet, whilst those of astronomers allow us to understand how stars and galaxies were formed and have evolved, how the elements came to have the relative abundances that we measure, and even to make a theory of how the universe began with a hot Big Bang.

Similarly, through an extension of the process of seeing familiar objects and deducing their properties, we have come to know of atoms and their nuclei, of protons and electrons, of neutrinos and quarks and the other particles of modern physics. We even have some understanding of how and why these objects behave in the way that they do.

All these pictures, of course, are initially inventions of the human mind. This means that they could be wrong; we could have misread the clues. Indeed, surely, some of our accepted notions of history, some of our ideas about the early stages of the universe, some of our theories regarding elementary particles, will almost certainly be false. The possibility that we can make mistakes, however, does not deny the existence of the truth, any more than

† Cowper, *Olney Hymns*, 35.

my failure to obtain the correct answer to a sum would deny the existence of such an answer. Indeed the possibility of being wrong itself suggests the existence of a truth. Although what we know of history is a human construction from the available evidence, it is a fact that somebody wrote the plays ascribed to Shakespeare, that some process was responsible for the planets, and that there was some state of the universe when the elements were formed. Equally, long before the human brain gave names to their constituents, there were protons and neutrons binding together to form nuclei. The way we describe reality is dependent upon the human brain and is therefore subject to the brain's limitations; it is an unnecessary arrogance on our part to assume that the reality itself is subject to similar limitations.

We have seen in this book that, at the level of quantum phenomena, the clues which we receive from our observations of the world become strange and even, from the point of view of reasonable pictures of reality, contradictory. How should we react to these discoveries?

Hidden-variable theories, involving the wavefunction as a pilot wave, are the only theories available which can make reasonable claims to be complete theories of quantum physics. They are complete in the sense that they do not require us to make any assumptions that certain things happen due to processes that are not described by the theory. The non-locality of the quantum world is explicit in these theories, and they are therefore very different from classical physics. The present forms of such theories are not convincing, and it would perhaps be very strange if quantum phenomena, which are often very beautiful, had their origin in theories that are decidedly inelegant. Such arguments, however, are very weak and depend upon subjective judgments, so we should be wary of using them, especially when the alternatives all involve a great deal of vagueness.

It may nevertheless be the case that, in accepting hidden-variable theories, we are ignoring the clues that quantum phenomena are giving us. These phenomena may be telling us that radically new ways of picturing reality are required. In one sense this was the message, the positive message, of the 'Copenhagen' interpretation: classical ways of thinking are no longer adequate, we need new ways of describing reality. The fact that it proved difficult to find such ways led to the negative aspect, namely that the quest was futile. This I do not believe.

Perhaps the most radical of all options is that of the many-worlds interpretation. This shares with hidden-variable theories the property of not requiring wavefunction reduction. The universe changes with time only through the deterministic, local, equation of ordinary quantum theory (Schrödinger's equation), and there is no need to have non-local interactions. All parts of the wavefunction exist for all time; none is selected through a non-deterministic process. Could it be that the reluctance of physicists to accept this idea, and to recognise the existence of a multitude of universes, is of a similar nature to the earlier reluctance to accept that our world is not the unique centre of all things?

Quantum phenomena provide us with many wonderful insights into the world. It could be that through some type of hidden-variable theory, or through developments along the lines of the Pearle corrections to quantum theory, we already have the beginnings of an explanation. If this is so then we must come to understand better the nature of the non-locality that is explicit in these theories. On the other hand, we may have to move further from the conventions of classical physics, to free our minds from many of the inbuilt prejudices of what can be discussed in physics, to be released from our restraints regarding time and space, . . . , in order to find a true explanation.

It is, in any case, an exciting prospect for the future.

Appendix 1

The Potential Barrier in Classical Mechanics

The easiest way to calculate the interaction of a particle with a potential barrier according to classical mechanics is to use conservation of energy.

The energy consists of two parts. The kinetic energy, which is due to the motion, is given by

$$\text{KE} = \tfrac{1}{2} m u^2 \tag{A1.1}$$

for a particle of mass m moving with velocity of magnitude u.

If, instead of a particle, we consider a rolling sphere then there is also a contribution to the kinetic energy due to rotation of the ball. In this case we have

$$\text{KE} = \tfrac{7}{10} m u^2. \tag{A1.2}$$

The potential energy, due to the gravitational attraction of the earth, is given by

$$\text{PE} = mgH + \text{constant} \tag{A1.3}$$

where g is the acceleration due to gravity, H is the height above some arbitrarily chosen base, and the constant can be any number we choose. The actual measured value of g at the earth's surface is

$$g = 9.81 \text{ m s}^{-2}. \tag{A1.4}$$

Thus the equation expressing conservation of energy

$$\text{KE} + \text{PE} = \text{constant} \tag{A1.5}$$

becomes, for non-rotating particles,

$$\tfrac{1}{2} m u^2 + mgH = \text{constant}. \tag{A1.6}$$

We can evaluate the constant using the initial conditions in which $H = 0$ and u takes the initial value which we denoted by v (see figure A1). Then at any later time we have

$$\tfrac{1}{2}mu^2 + mgH = \tfrac{1}{2}mv^2. \tag{A1.7}$$

The masses cancel leaving the equation
$$u^2 = v^2 - 2gH. \tag{A1.8}$$

We see from this equation that as H increases u decreases, i.e. the particle slows down as it climbs the hill. One of two things can now happen. Either H increases until u becomes zero, or the particle reaches the top of the hill, $H = H_T$, before u has become zero. In the former case the particle rolls back down again, whilst in the latter it continues over the other side, or in other words it is transmitted.

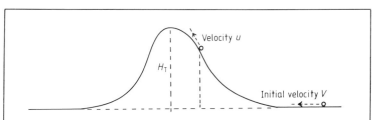

Figure A1 The potential barrier calculation. At a height given by H the particle has a velocity u.

The critical velocity, V, is therefore given by the value of v for which the right-hand side of equation (A1.8) becomes zero for H equal to H_T. Thus,

$$V^2 = 2gH_T. \tag{A1.9}$$

Whenever v is greater than V, the particle will be transmitted by the hill, whenever it is less then the particle will be reflected.

For rolling spheres it is easy to see that the corresponding critical velocity is given by

$$V^2 = \tfrac{10}{7}gH_T. \tag{A1.10}$$

Appendix 2

A Romantic Interlude

It is 16:30 in the first-year physics laboratory of the University of Somewhere. A lone student, Paul, is completing an experiment in which he records the times when clicks occur in a Geiger counter. These clicks signify the decay of particular nuclei in a radioactive substance, decays which, being quantum phenomena, happen at random times. The student wishes to study various aspects of probability distriøutions and, to this end, he has been told to record the exact times of 50 clicks. We shall suppose that the average period of time for these is 25 minutes. However, it could only take 24 minutes, say, or equally it could take 26 minutes.

In our first scenario, then, the time taken is 24 minutes. The student completes his record, switches everything off, thinks the experiment pretty dull, collects all his belongings and leaves. That night, when perhaps he should have been analysing his results, he goes rugby training.

Things, however, could have been very different. One atomic nucleus, recorded above as having decayed, might instead have chosen not to decay. Thus Paul has to wait for 26 minutes in order to complete his experiment. Just as he is about to leave the laboratory, Judy, another first-year physics student, comes running in. She had left her pen on the desk earlier in the afternoon. Paul has seen it and he tells her where she can find it. They talk and introduce themselves, they walk back to their rooms together, they arrange to meet later that evening, Paul misses rugby training and, in consequence, is not picked for the team on the following Saturday ...

Readers are invited to continue these stories as they wish. Clearly the two stories may diverge by arbitrarily large amounts. Microscopic effects can have macroscopic consequences.

Appendix 3

The Probability Function

The probability function, $P(x)$, for one particle moving in one space dimension is defined so that the probability of finding the particle in any small interval is given by the product of P, evaluated at the interval, and the length of the interval. Thus, if we consider the interval between the points labelled x and $x + dx$, where dx is a small length, then this probability is $P(x)\,dx$. As we see in figure A2, this quantity is the area of the small, approximately rectangular, column, standing on the interval.

It follows that the probability of finding the particle between the points labelled a and b is given by

$$P(a, b) = \int_a^b P(x)\,dx \qquad (A3.1)$$

which is the area bounded by the curve $P(x)$ and the lines $x = a$ and $x = b$.

The probability that the particle is somewhere on the line, i.e. that it exists, is equal to unity. Hence we have

$$\int_{-\infty}^{+\infty} P(x)\,dx = 1. \qquad (A3.2)$$

For a particle moving in three space dimensions these expressions are readily generalised. Instead of a small line with length dx, we consider a small rectangular volume with corners given by (x, y, z), $(x + dx, y, z)$, $(x, y + dy, z)$, $(x, y, z + dz)$ etc, as shown in figure A3. This has volume $dx\,dy\,dz$. The probability of finding the particle inside this small volume is given by the product of the volume and a probability function evaluated at the position of the volume, i.e. $P(x, y, z) \, dx \, dy \, dz$. The fact that the particle is somewhere is given by the equation, analogous to equation (A3.2),

$$\int_{-\infty}^{+\infty} dx \int_{-\infty}^{+\infty} dy \int_{-\infty}^{+\infty} dz \, P(x, y, z) = 1. \qquad (A3.3)$$

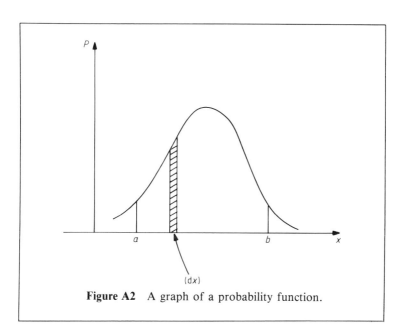

Figure A2　A graph of a probability function.

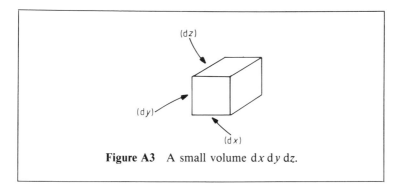

Figure A3　A small volume $dx \, dy \, dz$.

Appendix 4

The Wavefunction

We shall here denote the wavefunction by the letter W. For a single particle moving in one space dimension, W will depend upon the position on the line, that is upon some distance x. We can indicate this dependence by writing the wavefunction as $W(x)$.

The fact that W, at each point x, is actually a line, as explained in §2.2, is normally expressed by saying that W is a complex number. The length of the line, which in the text we call the magnitude of the wavefunction, is the *modulus* of the complex number and the angle is its *phase*. We can write W in the form

$$W = |W| e^{i\theta} \tag{A4.1}$$

where $|W|$ is the modulus and θ is the phase. Both these quantities will, in general, vary with x.

Any complex number can be written as the sum of a 'real' part (which is defined to be the projection of the line along a fixed direction, as shown in figure 7), and an 'imaginary' part (which is the projection of the line along the perpendicular direction) multiplied by the purely imaginary number i which is the square root of minus one ($i^2 = -1$). Readers should realise that the terms real and imaginary, as used here, refer simply to the projections of the line along two particular directions. *No connection with our other use of the term real, as in 'reality', etc, should be made.*

The probability function of Appendix 3 is related to $W(x)$ by

$$P(x) = |W(x)|^2. \tag{A4.2}$$

Thus, for a particle that is known to be in some small region around a particular point, e.g. with a $P(x)$ like that of figure 5, the modulus of $W(x)$ peaks around that point and is zero far from it. The condition that we have one particle, equation (A3.2), now becomes

$$\int_{-\infty}^{+\infty} W(x)\,\mathrm{d}x = 1. \tag{A4.3}$$

In the limiting case, where the particle has an exact position, the width of the peak becomes zero and its height infinite. The function $P(x) = |W(x)|^2$ is then called the *Dirac delta function*.

The opposite situation is when a particle has a fixed velocity. Here we have

$$W(x) = N\mathrm{e}^{\mathrm{i}kx} \tag{A4.4}$$

where N is a constant and the velocity is related to k by

$$v = \hbar k/m \tag{A4.5}$$

with m the mass of the particle. For this wavefunction

$$|W(x)|^2 = N^2 \tag{A4.6}$$

which is independent of x. The particle therefore has equal probability of being at all points of space. This is in accordance with the uncertainty principle of §2.2; when the uncertainty in velocity is zero, that in position has to be infinite.

Using the fact that

$$\mathrm{e}^{\mathrm{i}kx} = \cos kx + \mathrm{i}\sin kx \tag{A4.7}$$

we see that the real and imaginary parts of the wavefunction in equation (A4.4) look like ordinary waves, e.g. as in figure A4. In a more general situation like that of figure 5, for example, the real and imaginary parts form a 'wave packet' as in figure 8.

With a wavefunction like that given by equation (A4.4) there is clearly a difficulty in satisfying the normalisation condition given in equation (A4.3). In fact, since $|W(x)|^2$ is constant the only value of N consistent with this condition is $N = 0$. We can easily understand this: if a particle has equal probabilities for being anywhere along an infinite line, then its probability of being in any finite region is clearly zero. This is a technical problem which can easily be overcome, for example, by putting the whole universe in a finite box. In practice, of course, a particle which is of interest to us must lie within some finite distance from our experiment, so this restriction to a finite box will not have any significant effect.

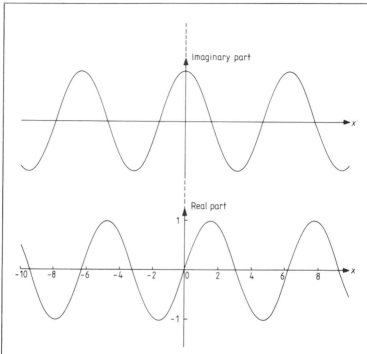

Figure A4 Showing the real and imaginary parts of the wavefunction e^{ikx}. This represents a particle with precise velocity ($v = \hbar k/m$) and totally uncertain position.

We shall now use this wavefunction, corresponding to precise velocity, to do a small calculation. We shall calculate, in a very approximate way, the probability of transmission through a potential barrier as described in §1.3 and in Appendix 1. The argument is as follows. We suppose that the initial velocity is less than the critical velocity (V); then in the region where, according to classical mechanics, the particle cannot exist, that is where

$$H > v^2/2g \qquad (A4.8)$$

the velocity becomes purely imaginary. In fact, as we see from equation (A1.8) it is given by

$$u = \mathrm{i}(2gH - v^2)^{\frac{1}{2}}. \tag{A4.9}$$

The wavefunction (A4.4) becomes

$$W = \exp[-(2gH - v^2)^{\frac{1}{2}}xm/\hbar] \tag{A4.10}$$

which is a *decaying* rather than oscillating function. Thus, as the wavefunction leaks into the classically forbidden region, it reduces exponentially. The probability of transmission depends upon how much the wavefunction has decreased by the time it reaches the classically allowed region on the other side of the barrier (see figure A5). Roughly speaking this probability is equal to W as given in equation (A4.10), with x being the distance the particle has to travel in the forbidden region, i.e. d in the figure.

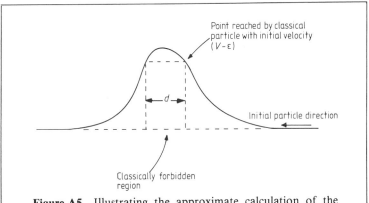

Figure A5 Illustrating the approximate calculation of the transmission probability when the velocity is such that the particle would be reflected according to classical mechanics. In the classically forbidden region the wavefunction is approximately $\exp[-(2gH_{\mathrm{T}} - (V - \varepsilon)^2)^{\frac{1}{2}}xm/\hbar]$.

We shall evaluate this for numbers relevant to the normal situation in macroscopic physics. Thus we take a mass of 10 g, and suppose that the critical velocity is 10 cm s^{-1} and the actual initial velocity is $V - \varepsilon$, with $\varepsilon = 10^{-5}$ cm s^{-1} (corresponding to a departure from the critical velocity of one part in 10^6). Then $2gH - v^2$ is approximately equal to $2V\varepsilon$ and the coefficient in the exponential of equation (A4.10) becomes $(2V\varepsilon)^{1/2}md/\hbar$. If the forbidden region is about 1 cm long this gives a probability of transmission equal to about $\exp(-10^{26})$, which is sufficiently near to zero to justify the remark made in §1.3, that departures from the causal predictions of classical mechanics will not be seen in macroscopic physics.

We return now to our general discussion of the wave function. In order to obtain the probability of finding the particle with a particular velocity, when the wavefunction is not of the simple form given in equation (A4.4), we use a mathematical theorem that allows us to write any (reasonable) wavefunction as an integral over wavefunctions with fixed velocity. Thus we can put

$$W(x) = \int_{-\infty}^{+\infty} e^{ikx}\, \overline{W}(k)\, dk \qquad (A4.11)$$

where $\overline{W}(k)$ is a new function defined by this equation. Then the probability of finding a particular value of k is given by $|W(k)|^2$. In other words $\overline{W}(k)$ is related to the values of k in the same way that $W(x)$ is to the values of x.

In §2.2 we defined the addition of two wavefunctions. The rule is the same as that for the addition of two complex numbers. Then, if W and V are two wavefunctions, we have

$$|W + V|^2 = |W|^2 + |V|^2 + W^*V + WV^*. \qquad (A4.12)$$

The last two terms are those which are responsible for interference effects.

Now we must consider the situation when we have two particles, which we call A and B, and which for simplicity we confine to a line. The wavefunction is now a function of two positions, x_A and x_B, so we write it as $W(x_A, x_B)$. The probability of finding particle A in the region x_A to $x_A + dx_A$ and the particle B in the region x_B to $x_B + dx_B$ is given by

$$P(x_A, x_B)\, dx_A\, dx_B = |W(x_A, x_B)|^2\, dx_A\, dx_B. \qquad (A4.13)$$

By integrating over a region of x_B, for example, we can now find the probability function for A given that B is in this region. Thus, if B is in the region $x_1 < x_B < x_2$, then the probability distribution for A is given by

$$P(x_A) = \int_{x_2}^{x_1} |W(x_A, x_B)|^2 \, dx_B. \qquad (A4.14)$$

This distribution will, in general, depend on the region chosen for x_B, i.e. on where B is. This implies that there are correlations in the positions of the two particles, or in other words that the probability of finding one of them at a particular place depends upon the position of the other. Such correlations would be expected if there are forces between the two particles; for example, if the particles attract each other then they are likely to be found close together. The exception to this occurs when the wavefunction, which is a function of two variables, can be written as the product of two functions each of one variable:

$$W(x_A, x_B) = U(x_A)V(x_B). \qquad (A4.15)$$

In this case it is easy to see from equation (A4.14) that the distribution for A, for example, is totally independent of the position of B. This factorisation property sometimes provides a useful approximation, but it never provides a true description. In the real world particles are always correlated, so the wavefunction is a genuine function of two, or in the case of N particles N, variables, which cannot be written as the product of functions of one variable.

One important source of correlation between identical particles is the quantum theoretical requirement that the wavefunction must be either totally symmetrical, or totally antisymmetrical, when the particles are interchanged. Thus if we consider two identical particles with positions denoted by x_1 and x_2, respectively, the wavefunction must satisfy either

$$W(x_1, x_2) = \pm W(x_2, x_1) \qquad (A4.16)$$

with the $+$ sign giving symmetry and the $-$ sign giving antisymmetry. A product wavefunction of the form $U(x_1)V(x_2)$ does not satisfy either of these conditions (unless $U = V$). In order to antisymmetrise (for example) such a wavefunction it must be replaced by

$$W = U(x_1)V(x_2) - U(x_2)V(x_1). \qquad (A4.17)$$

This last equation shows an important effect caused by anti-symmetry. If U and V are the same wavefunction, then W becomes zero. Thus there is no possibility of having two particles, with an antisymmetric wavefunction, in the same state. This fact is known as the Pauli exclusion principle. It explains why the electrons in atoms are not all in the same states, and hence why we have such a rich variety of chemical properties.

Appendix 5

The Hydrogen Atom

A hydrogen atom consists of a negatively charged electron and a positively charged proton. Since the proton is about 2000 times heavier than the electron we can regard it as stationary, with the electron moving around it. The force between the two charges is their product divided by the distance between them squared, i.e.

$$F = e^2/r^2 \qquad (A5.1)$$

where we have used the symbols e for the magnitudes of the two charges, which are equal apart from the sign, and r for the separation. The direction of the force is along the line joining the particles, and it is attractive. Some readers may be accustomed to seeing equation (A5.1) with various extra factors; we have chosen to measure charge in such units that these factors are not required. According to classical mechanics, the electron would follow a circular or elliptical orbit around the proton, just like a planet around the sun. All such orbits would be possible solutions, and they would be very unstable, due to collisions and the radiation of light waves, which is contrary to what is observed.

The situation in quantum theory is very different, however, because there is a wavelength associated with the electron. According to equation (2.4) this is given by

$$l = h/mv \qquad (A5.2)$$

where m is the mass of the electron and v is its velocity. The quantum condition now arises because the circumference of the orbit has to contain exactly an integral multiple of wavelengths, otherwise, having gone all the way around, the wave would not be back to its initial value. Thus we require

$$nl = L \qquad n = 1, 2, 3, \ldots \qquad (A5.3)$$

where L is the circumference (see figure A6).

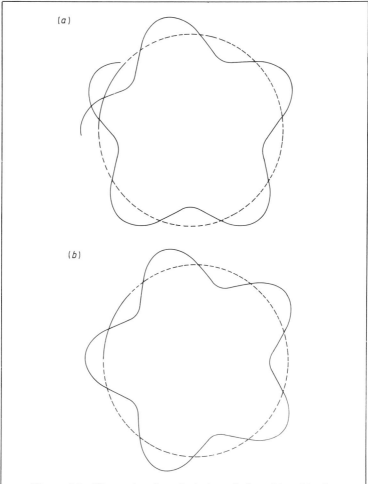

Figure A6 Illustrating the calculation of allowable orbits for the hydrogen atom. The orbit in (*a*) does not close on itself and is therefore not allowed. The orbit in (*b*) is allowed.

We now restrict our attention to circular orbits (this does not give all possibilities but it gives the lowest, normal state of the atom). In this case we have that

$$L = 2\pi r. \tag{A5.4}$$

Also the balance between the centrifugal force and the electric force of equation (A5.1) gives

$$mv^2/r = e^2/r^2 \tag{A5.5}$$

or

$$v^2 = e^2/mr. \tag{A5.6}$$

Putting this into equation (A5.2), and using the equations (A5.3) and (A5.4), yields immediately

$$r = \hbar^2 n^2/e^2 m. \tag{A5.7}$$

This gives the radii of all possible circular orbits. The smallest, with n equal to one, is the radius of the hydrogen atom in its lowest energy state. The energy of the states can easily be calculated from the above equations and we find

$$E_n = -me^4/2n^2\hbar^2 \tag{A5.8}$$

a formula which has been verified to high accuracy.

This calculation, which we have based on a rather pictorial argument, can be done using the Schrödinger equation and the same energy level formula is obtained for the rotationally symmetric states (i.e. the circular orbits).

Appendix 6

Interference, Macroscopic States and Orthogonality

We suppose that we wish to measure a particular quantity, M, for a system described by a wavefunction $W(x)$. Here the x is used to represent all the variables in the problem, e.g. just a simple distance for one particle moving on a line, or, more generally, several positions for a set of particles moving in space. The measurements can give one of a set of values m_i, with $i = 1, 2, 3$, etc. Here we are assuming that the results of the measurement are discrete; the extension to the case where they lie in a continuum can easily be made. Corresponding to each i there will be a wavefunction $w_i(x)$, such that if the system is in the state with wavefunction $w_i(x)$, then the measurement would always give the result m_i. It is convenient to assume that the wavefunctions, $w_i(x)$, are all normalised to unity, i.e. that the square of their modulus, integrated over all values of the variables, x, is one.

It can be shown that it is always possible to express any wavefunction $W(x)$ as a sum of the $w_i(x)$, with particular constant coefficients. That is, we can write

$$W(x) = c_1 w_1(x) + c_2 w_2(x) + c_3 w_3(x) + \dots \qquad \text{(A6.1)}$$

where the c_i are unique complex numbers, which do not depend upon the x. These numbers are in fact related to the probability for finding the result m_i, when M is measured for the system. This probability is given by $|c_i|^2$.

Now let us include the measuring apparatus into our discussion. Corresponding to each result, m_i, this will have a state with wavefunction $U_i(y)$, with $i = 1, 2, 3$, etc, and where we have denoted the variables associated with the measuring apparatus by y. Again the $U_i(y)$ are taken to be normalised to unity. Assuming that the apparatus is suitable for its purpose, we will be able to 'see' in which state it is, and so will know the result of the measurement.

The complete wavefunction of the original system plus the measuring apparatus, after they have interacted, i.e. after the measurement has been made, will have the form:

$$V(x, y) = c_1 w_1(x)U_1(y) + c_2 w_2(x)U_2(y)$$
$$+ c_3 w_3(x)U_3(y) + \ldots \qquad (A6.2)$$

Note that the measuring apparatus 'works' correctly in that observation of the apparatus to be in the state U_1, for example, selects that part of the wavefunction corresponding to the system being in the state w_1.

On the other hand, our apparatus has not itself 'selected' one part of the wavefunction, i.e. the wavefunction has not been reduced. This is in accordance with our discussion of §3.5. The point we want to consider now is whether there are any experimental consequences of this fact. In other words we wish to know whether we can *see* the difference between the wavefunction of equation (A6.2) and the corresponding reduced wavefunction:

$$W = w_1(x)U_1(y) \qquad \text{with probability } |c_1|^2$$
$$\text{or } w_2(x)U_2(y) \qquad \text{with probability } |c_2|^2 \qquad (A6.3)$$

etc.

In order to do this we might, for example, calculate the average value of x at some later time of the experiment. This is given by the general expression

$$x_{av} = \iint W^*W \, dx \, dy. \qquad (A6.4)$$

If we use equation (A6.3) for W this becomes

$$x_{av} = \int |c_1|^2 x |w_1|^2 \, dx + \int |c_2|^2 x |w_2|^2 \, dx + \ldots \quad (A6.5)$$

where we have used the fact that the wavefunctions for the apparatus are properly normalised, i.e.

$$\int |U(y)|^2 \, dy = 1. \qquad (A6.6)$$

The result obtained with the unreduced wavefunction, equation (A6.2), is the same as this except for the addition of 'cross terms' which have the form

$$\int\int x c_i^* \, w_i(x)^* U_i(y)^* c_j w_j(x) U_j(y) \, \mathrm{d}x \, \mathrm{d}y$$

with i and j different. However, it is a mathematical theorem that the y integrals in expressions of this type are always zero. This theorem is expressed by the statement that different values for measured quantities are associated with wavefunctions that are mutually orthogonal. Thus the fact that our original system has interacted with the measuring apparatus means that all the cross terms, which are the ones responsible for interference effects, disappear. As we stated in §3.6, it is necessary that the unobserved parts of the system are in exactly the same state if interference effects are to be seen.

Appendix 7

A Gambling Interlude

In this appendix we shall discuss further how it is possible to modify the rules of quantum theory in order to permit explicit reduction of the wavefunction, as explained in §3.7. The ideas are due mainly to Pearle (see Pearle's article, 'Dynamics of the reduction of the state vector', in *The Wave Particle Dualism* edited by Diner *et al* [Dordrecht: Reidel 1984]).

We consider a complete system, i.e. including any measuring apparatus, and describe it by a wavefunction $W(t)$. Here we have not explicitly indicated the variables on which the wavefunction depends since they play no role in our discussion. We have, however, put in the time variable, t, as we are concerned with the way W varies with time.

We now choose some particular observable, associated with the system. Corresponding to each value of this observable there will be a wavefunction for the system. We denote the set of such wavefunctions by w_i, with $i = 1, 2, 3$, etc. As we noted already in Appendix 6, it is always possible to write any W as a sum over the w_i:

$$W(t) = c_1(t)w_1 + c_2(t)w_2 + \ldots \qquad (A7.1)$$

The problem now is to devise an equation for the time dependence of W such that the following properties hold:

(i) The normalisation is preserved, i.e.

$$|c_1(t)|^2 + |c_2(t)|^2 + \ldots = 1 \qquad (A7.2)$$

for all times.

(ii) All $c_i(t)$ go to zero as t becomes large, except for one of them which, from equation (A7.2), must become of magnitude one.

(iii) Given the values of the $c(t)$ at a time t, the probability that at a much later time $t = T$ (T very large), a particular c_i will have become unity, i.e. $|c_i(T)| = 1$, is proportional to $|c_i(t)|^2$.

The first of these properties is, of course, satisfied by the Schrödinger equation. The others are not. The second property states that, after a sufficiently long time, wavefunction reduction will have taken place. The last property ensures that the predictions of the modified theory agree with those of quantum mechanics, namely that the probability of reduction to the state denoted by w_i is equal to the probability of observing the value corresponding to this state.

It is clear that some random element is involved with the last property—it can never arise if the time behaviour of the c_i is deterministic. This randomness is introduced by Pearle through the addition of an extra, non-linear, term into the Schrödinger equation, a term which contains a random input. Thus the theory involves 'stochastic' differential equations. We shall not enter this area but, instead, will discuss a very simple 'model' which has the above three properties.

We consider two gamblers, Charlie and Wally, who are playing a game with each other. Let $X_C(t)$ and $X_W(t)$ be the amount of cash held by Charlie and Wally, respectively, at time t. Then we know that

$$X_C(t) + X_W(t) = 1 \qquad (A7.3)$$

where we have called the total amount of money 1, as is always possible by suitable choice of our unit.

They now play a game of chance and agree that if Charlie wins he receives 0.01 from Wally, and vice versa. They play until eventually one or the other has no money left so that

$$\text{either } X_C = 1 \text{ and } X_W = 0$$

$$\text{or} \qquad X_C = 0 \text{ and } X_W = 1 . \qquad (A7.4)$$

If we identify the X with the previous $|c|^2$, then the gambling model clearly satisfies the first two of the required properties for our modified quantum theory. (We only have a two-state system of course, but it is easy to generalise by adding more people to the game.)

It turns out that the third property is also true in the model. To demonstrate this we denote the probability for Charlie to win eventually by $P(X_C)$. We expect this probability to depend upon the amount of money he has; thus, as indicated, it is a function of X_C. Now consider the situation after one game has been played. Assuming that the games are fair, there is a 50% chance that Charlie will have $X_C + 0.01$, and a similar chance that he will have $X_C - 0.01$. Provided we include both these possibilities the probability of his winning eventually cannot have changed, so

$$P(X_C) = \tfrac{1}{2}P(X_C + 0.01) + \tfrac{1}{2}P(X_C - 0.01). \qquad \text{(A7.5)}$$

We also know that $P(0) = 0$ (he has already lost) and $P(1) = 1$ (he has won). Clearly all these equations are satisfied by

$$P(X_C) = X_C \qquad \text{(A7.6)}$$

which can also be shown to be the only solution. This equation is the required third condition.

Although we have here a very pretty model we remind ourselves that it is unsatisfactory as a solution to the problem of wave-function reduction. There are two reasons for this. First, it was necessary to choose a particular observable in order to specify the states w_i. Clearly a different choice would cause reduction to a different set of states. How does a system know to what set of states it should reduce? Or, in other words, how does the system know what is going to be observed? The second problem concerns the time scale over which reduction takes place, i.e. how large does the T have to be? In some cases it has to be very small, whereas we know that in others there is no reduction, i.e. no deviation from orthodox quantum theory, over enormous times.

There are possible ways around these difficulties. It is reasonable to argue that, ultimately, all measurements are position measurements, e.g. the position of a pointer on a dial, or even the ink in some computer output, so the preferred states to which all states finally reduce might well be states corresponding to unique, or at least almost unique, positions.

The answer to the second problem could be that the rate at which the reduction happens depends on the mass of the object being considered. One possibility would arise if we suppose that the reduction process has to conserve the average value of the energy. Then, if we consider a state with unique energy E reducing to a wavepacket of size L, we have

$$E = \hbar^2/2mL^2 \qquad (A7.7)$$

where we have used the uncertainty relation, equation (2.1), to relate the size of the wave packet to the average of the magnitude of the momentum. Now the maximum energy of a non-relativistic particle of mass m is certainly less than mc^2, where c is the velocity of light. So equation (A7.7) gives

$$L > \hbar/mc. \qquad (A7.8)$$

For an electron this is about 10^{-10} cm, which is already quite small, though much bigger than the radius of an electron, which is less than 10^{-18} cm. For a macroscopic object of mass about 1 g, the limit on L from equation (A7.8) is 10^{-38} cm, so reduction to a unique position would be allowed to essentially perfect accuracy.

Another possible way of guessing the effect of the mass would be to suppose that in some way gravity is responsible for wavefunction reduction. (See §6.2 for some reasons why we might want to believe this.) Then to estimate the time required for the reduction to occur we would have to construct a time out of the only available constants, $\hbar$, m and G, the constant of gravity. This can be done in only one way, namely by the combination

$$T = \hbar^3/G^2m^5. \qquad (A7.9)$$

For an electron this is 10^{70} secs, whereas for a particle of mass 1 g it is 10^{-67} s. If we could find a theory in which equation (A7.9) gave the time for reduction to occur, then for elementary particles it would essentially never occur, whilst for macroscopic objects, it would be instantaneous. Such a theory would be very satisfactory. We stress however, that, at the present time, it does not exist.

It could be that explanations along these lines can be constructed. They will certainly involve very strange behaviour. To emphasise this fact we consider our potential barrier experiment. In any explanation along the lines of this appendix, we would have to believe that the two peaks in the wavefunction, which give the probabilities of reflection and transmission (see figure 11), are continuously interchanging small amounts in a random fashion. Such interchanges happen regardless of how far apart the peaks are, and occur more rapidly the greater the number of particles involved. Things would be 'going on' in the world that are very contrary to our usual experience. Where would we look to see other, more direct, evidence of these phenomena?

Appendix 8

Spin

In classical mechanics the 'spin' of an object is its angular momentum about its centre of mass. It is a vector quantity, having a magnitude and a direction. An example is given in figure A7 where the length of the line represents the magnitude of the spin and, of course, the direction of the line is the direction of the spin.

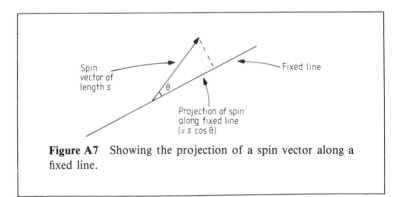

Figure A7 Showing the projection of a spin vector along a fixed line.

The *component* of the spin about any fixed line is the projection of the spin along that line as indicated in the figure. Thus, from elementary trigonometry, if s is the magnitude of the spin, then its component along the line denoted by **1** is

$$s_1 = s \cos \theta \qquad (A8.1)$$

where θ is the angle between the line **1** and the spin direction (see figure A7).

The spin, s, in classical mechanics, can take any value, and its projection onto a given line, according to equation (A8.1), can then lie between $-s$ and $+s$, the limiting values occurring when the direction is either parallel or antiparallel to the line.

Now we turn to quantum mechanics. Here the magnitude of the spin can only be one of the values $[n(n+1)]^{1/2}\hbar$, with n equal to $0, \frac{1}{2}, 1, \frac{3}{2}, 2, \frac{5}{2}$, etc. We refer to these values as corresponding to spins $0, \frac{1}{2}, 1$, etc. Further, the component of a spin n particle in any given direction will always have one of the values $-n$, $-n+1$, $-n+2$, ..., $n-2$, $n-1$ or n. For example, with spin zero the component will always be zero, with spin $\frac{1}{2}$ it will be $+$ or $-\frac{1}{2}$, with spin 1 it will be $+$ or -1, or zero.

For simplicity we shall now restrict ourselves to the simplest non-trivial case, namely spin $\frac{1}{2}$. Suppose we have a spin $\frac{1}{2}$ particle and we know that its spin component along the line **1**, which we write as s_1, is $+\frac{1}{2}$. If we now measure the spin component, s_2, along a new direction, denoted by **2**, then we will find a value equal to either $+\frac{1}{2}$ or $-\frac{1}{2}$. Does quantum theory tell us which value we will obtain? As we might expect it does not; rather it tells us the *probability* of getting a particular value.

A proper calculation of this probability would take us further into the formalism of quantum theory than we can go here. However, there is a simple way of obtaining the result if we assume, correctly, that the probability will be such that the average value of the result will be equal to the value predicted by classical mechanics, i.e.

$$s_2^{\text{av}} = s_2^{\text{class}} = \tfrac{1}{2}\cos\theta \qquad (\text{A8.2})$$

where θ is the angle between the lines **1** and **2**.

Now we let $P(+, +)$ be the probability of obtaining $s_2 = \frac{1}{2}$ when $s_1 = \frac{1}{2}$, and $P(-, +)$ be the corresponding probability of obtaining $s_2 = -\frac{1}{2}$. Then

$$s_2^{\text{av}} = \tfrac{1}{2}P(+, +) - \tfrac{1}{2}P(-, +). \qquad (\text{A8.3})$$

Also, since the probability of obtaining either $+\frac{1}{2}$ or $-\frac{1}{2}$ for s_2 must be one, we have

$$P(+, +) + P(-, +) = 1. \qquad (\text{A8.4})$$

Thus

$$P(+, +) = \tfrac{1}{2}(\cos \theta + 1) = \cos^2(\theta/2) \qquad \text{(A8.5)}$$

and

$$P(-, +) = \sin^2(\theta/2). \qquad \text{(A8.6)}$$

These are the required expressions.

Similarly, if we define $P(+/-, -)$ as the probability of obtaining $+/-\tfrac{1}{2}$ along the new direction when we have $-\tfrac{1}{2}$ along the original direction, we find

$$P(+, -) = \sin^2(\theta/2) \qquad \text{(A8.7)}$$

and

$$P(-, -) = \cos^2(\theta/2). \qquad \text{(A8.8)}$$

In general, to specify the spin state of any spin $\tfrac{1}{2}$ particle we must first choose a particular direction and then give the probability of the particle having spin $+$ or $-\tfrac{1}{2}$ in this direction. When we measure the spin in the new direction we choose one of these possible values and, at least with a suitable measuring apparatus, the particle will now have a definite spin in this direction. In all other directions, except the direction which is antiparallel, i.e. $\theta = 180°$, the spin will not be definite, as follows from equations (A8.5) to (A8.8).

Actually, probabilities are the moduli squared of 'probability amplitudes' (recall the formulae of Appendix 4), and it is these amplitudes that we shall need. Thus we put

$$P(+, +) = |A(+, +)|^2 \qquad \text{(A8.9)}$$

and

$$P(-, +) = |A(-, +)|^2 \qquad \text{(A8.10)}$$

etc. These equations and the previous equations for the P do not completely define the probability amplitudes. In fact the full theory gives

$$A(+, +) = A(-, -) = \cos(\theta/2) \qquad \text{(A8.11)}$$

$$A(-, +) = -A(+, -) = \sin(\theta/2). \qquad \text{(A8.12)}$$

In order to discuss the correlations in EPR-type experiments we now consider two spin $\tfrac{1}{2}$ particles, which we denote by A and B as in the text. The total spin is obtained by 'adding' the two spins and it turns out that, in quantum theory, two values of this total spin are possible, namely one and zero. It is not hard to understand this; it means that the spin projection of one along the spin direction of the other is either $+\frac{1}{2}$ or $-\frac{1}{2}$; in the former case the spins are parallel and the total spin is one, whereas in the latter they are antiparallel leading to a total spin of zero.

We are concerned in the text with a state of total spin zero. To describe this in terms of the two particles we first choose a direction. Whatever direction we choose the component of the total spin along it is zero; thus, if one particle has a component of $+\frac{1}{2}$ then the other must have $-\frac{1}{2}$, and vice versa. The wavefunction will therefore consist of a term $V^+ W^-$ together with a term $V^- W^+$. Here V and W refer to particles A and B respectively and the $+/-$ give the spin component along the chosen direction.

We now consider the possibility of measuring the spin components of A in a direction making an angle a with the direction chosen above for defining the state, and of B in a direction defined by the angle b (see figure 21 and figure A8). In table A8.1 we show the probability amplitude for obtaining various spin combinations, due to each of the two terms in the wavefunction.

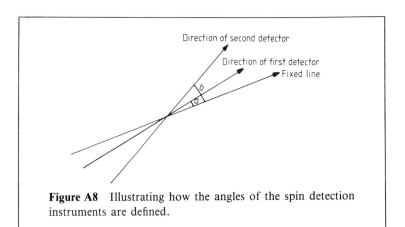

Figure A8 Illustrating how the angles of the spin detection instruments are defined.

The first use we make of this table is for the angles a and b being equal. In this case we require that the probability, and hence the probability amplitude, of the result $+, +$ or of $-, -$ is zero. Clearly this requires that we must subtract the two contributions in the table, i.e. that there has to be a relative minus sign between them. Since the overall sign of a wavefunction is irrelevant, this means that we can write for the complete wavefunction of the spin-zero state

$$Z = V^+ W^- - V^- W^+ \tag{A8.13}$$

as stated in the text. Note that although we have used a particular direction to define the states, V and W, of the two particles in equation (A8.13), the state itself is in fact independent of this direction; it is the unique state that has zero spin projection in *any* direction.

Next we consider the case where we measure the spins along different directions. In table A8.2 we show the probability amplitudes for obtaining various combinations of results, as calculated from the amplitudes given in table A8.1 and with the wavefunction of equation (A8.13). From this table we see that $+$, $+$ and $-$, $-$, both of which contribute $+1$ to E, have probability $\frac{1}{2}\sin^2[(a-b)/2]$, and that $+$, $-$ and $-$, $+$, which contribute -1 to E, have probability $\frac{1}{2}\cos^2[(a-b)/2]$. Note that we have here normalised these probabilities so that the probability of one of the four happening is unity. The quantity we require is thus given by

$$E^{\mathrm{av}}(a,b) = \sin^2[(a-b)/2] - \cos^2[(a-b)/2]$$
$$= -\cos(a-b) \tag{A8.14}$$

which is the result given in the text.

Table A8.1

A	B	$V^+ W^-$	$V^- W^+$
$+$	$+$	$-\cos(a/2)\sin(b/2)$	$-\sin(a/2)\cos(b/2)$
$-$	$-$	$\sin(a/2)\cos(b/2)$	$\cos(a/2)\sin(b/2)$
$+$	$-$	$\cos(a/2)\cos(b/2)$	$-\sin(a/2)\sin(b/2)$
$-$	$+$	$-\sin(a/2)\sin(b/2)$	$\cos(a/2)\cos(b/2)$

Table A8.2

Amplitude for

$+$, $+$ = $-\cos(a/2)\sin(b/2) + \sin(a/2)\cos(b/2) = \sin[(a-b)/2]$

$-$, $-$ = $\sin(a/2)\cos(b/2) - \cos(a/2)\sin(b/2) = \sin[(a-b)/2]$

$+$, $-$ = $\cos(a/2)\cos(b/2) + \sin(a/2)\sin(b/2) = \cos[(a-b)/2]$

$-$, $+$ = $\sin(a/2)\sin(b/2) + \cos(a/2)\cos(b/2) = \cos[(a-b)/2]$

Appendix 9

A Music Hall Interlude

Two music hall performers, Kevin and Margaret, claim to be able to demonstrate telepathy. Their act is as follows. The stage is divided into two sides by a screen and they go to separate sides. The audience is invited to verify that there can be no possible communication between them (by normal means).

A member of the audience is asked to write down a number: 1, 2 or 3. This is then given to a person on the stage who has in his possession three cards, numbered 1, 2 or 3, each bearing a question requiring a yes/no answer. The questions might, for example, be: 1. Have you ever been to Croydon?, 2. Do you like turnips? and 3. Do you like pop music? The appropriate card, bearing the number chosen, is then given to Margaret, who writes down her answer. Kevin sees and hears nothing of this.

Meanwhile, or before, or perhaps later, a similar procedure happens with Kevin. A random choice of 1, 2 or 3 is made from the same set of questions, and Kevin writes down his answer.

The questions and their answers are then shown to the audience. If it happens that the same question was selected then the audience notice that Margaret and Kevin gave the same answer.

The whole procedure is repeated several times, always with the same result: whenever the same question is selected, the same answer is given.

Kevin and Margaret perform their act for several nights. They move on to other towns, make an international tour, appear on many TV stations, become rich and famous, etc. Never once do they fail to give the same answer when the same question comes up.

According to their reaction to the performance, the audience can be divided into three groups. Some are out for a good evening's entertainment, are not too critical, clap loudly at everything, and, in particular, return home convinced that they have seen a good demonstration of telepathy (or some similar magic).

There are, of course, also the clever types who are not so easily taken in. 'Shame', they cry; 'Kevin and Margaret are frauds.' 'We can easily see how the trick is done. Before they go to the separate sides of the screen, Kevin and Margaret have agreed on their answers to questions 1, 2 and 3 (regardless of what the questions are). For example, they might have decided that to question 1 they will answer yes, to question 2 they will answer no, and to question 3 they will also answer no. Thus, while they wait for the question to be selected, they will merely have to remember their plan, YNN in the above example. Provided they do not deviate from this they are sure that the same question will receive the same answer.' 'It's easy.' 'How do they get away with such an obvious trick?' 'The audiences must be stupid.'

There may also be a few people who are even cleverer and who are also more observant. They realise that the method guessed at by the second group is the *only method* that will work, i.e. that will guarantee that in all cases the same question receives the same answer. It is *essential* that Kevin and Margaret should know the agreed answer to each question before they part. (To prove this we might suppose that Kevin did not know the answer to Q1. Then in the event that both had Q1 selected he would not know the answer given by Margaret, and would therefore only have a 50% chance of guessing the same answer.)

The third group of people have, however, also recorded the answers given by the two performers even when the questions were different. Over many months they have noticed that when all cases are considered (i.e. when the questions are ignored), Kevin and Margaret agree and disagree an equal number of times. This simple fact is sufficient to prove the existence of telepathy, or, at least, to show that there is some means of communication between Kevin and Margaret. Why?

We recall that we have shown that Kevin and Margaret must have some 'plan' for each trial. Suppose, for example, that the plan is YNN, as above, Let us then calculate the average number of agreements and disagreements with this plan. There are nine possible combinations of the three questions. These are shown in table A9.1, together with the answers given by Kevin and Margaret according to the plan. We see that agreement occurs in 5 cases and disagreement in 4. Clearly this is also true for any plan involving two Ys and one N, or two Ns and one Y. On the other hand, a plan involving three Ys or three Ns will lead to agreement in all cases, regardless of the questions. Thus we reach the conclusion that, whatever the plans chosen by Kevin and Margaret, there must be more agreements than disagreements. Since this is contrary to the observations, our assumption of no communication must be false. So long as we can assume that there is no trickery, Kevin and Margaret have provided conclusive and irrefutable evidence for telepathy.

Table A9.1

Questions asked: (Margaret first)	1,1	1,2	1,3	2,1	2,2	2,3	3,1	3,2	3,3
Margaret's answer:	Y	Y	Y	N	N	N	N	N	N
Kevin's answer:	Y	N	N	Y	N	N	Y	N	N
Agree/disagree:	A	D	D	D	A	A	D	A	A

(It is instructive to see how agreement could be obtained if communication were allowed. One possible way would be for Kevin to know the questions received by Margaret. Then, on the occasions when they both receive different questions, and when the plan tells him to give the same answer, he should sometimes, with a frequency that can be calculated, give the opposite answer.)

The story of Kevin and Margaret is, of course, a fable. I do not think it could happen. However, if we replace the two performers by electrons or photons, and the questions by projections of spin along suitably chosen directions labelled 1, 2 or 3, then we can be sure from the predictions of quantum theory, as verified by the Aspect *et al* experiments, that the performance would happen exactly as described. In this sense, 'telepathy' between particles is an established scientific fact. The only way of explaining the results without some form of communication is to accept the many-worlds interpretation.

Details of the physics involved for the set-up of this experiment, which requires three settings of the detectors rather than two as in §5.4, can be found in the excellent article 'Is the moon there when nobody looks?' by N D Mermin, *Physics Today*, April 1985. It was from this article that I learned the beautifully simple form of Bell's theorem implicit in the arguments of this appendix.

Index

Numbers in bold type refer to the start of relevant sections.

L.